11 SMART CITIES

11 SMART CITIES

Belinda Yuen, Yanjun Cai, Francine Chan,
Xin Yang & Kelly Lim

Singapore University of Technology and Design (SUTD), Singapore

World Scientific

NEW JERSEY • LONDON • SINGAPORE • BEIJING • SHANGHAI • HONG KONG • TAIPEI • CHENNAI • TOKYO

Published by

World Scientific Publishing Co. Pte. Ltd.
5 Toh Tuck Link, Singapore 596224
USA office: 27 Warren Street, Suite 401-402, Hackensack, NJ 07601
UK office: 57 Shelton Street, Covent Garden, London WC2H 9HE

British Library Cataloguing-in-Publication Data
A catalogue record for this book is available from the British Library.

11 SMART CITIES

ISBN 978-981-12-9573-7 (hardcover)
ISBN 978-981-12-9574-4 (ebook for institutions)
ISBN 978-981-12-9575-1 (ebook for individuals)

For any available supplementary material, please visit
https://www.worldscientific.com/worldscibooks/10.1142/13917#t=suppl

Desk Editor: Venkatesh Sandhya

ACKNOWLEDGEMENTS

Many people have contributed to the production of this book. We are grateful to our colleagues for their assistance and contributions:

- Darren Ho helped with preliminary literature review on Milan, Copenhagen, and Gothenburg.
- Nurhazimah Naharudin helped with preliminary literature review on Vancouver, Tokyo, and Auckland and data collection in Vancouver.
- David Hong helped with data collection in Boston.
- Emma Goh helped with literature review, data analysis, and write-up on Tokyo.
- Dr Hyunjoo Eom helped with data collection in Seoul and write-up on Seoul and Copenhagen.

A big thank you to the many people who patiently answered our interview questions and helped with our information requests.

CONTENTS

CHAPTER 1

INTRODUCTION

In recent decades, smart cities have proliferated and emerged as a global phenomenon that address urban challenges through innovative and intelligent solutions.[1] The development of smart cities is adjudged as an essential strategy towards Sustainable Development Goal 11: inclusive, safe, resilient, and sustainable cities.[2] Interest in smart cities and their effect on sustainable development has grown, including studies to review and rank smart cities worldwide on various key dimensions, e.g., the SUTD-IMD Smart City Index[3] and the City in Motion Index.[4] The SUTD-IMD Smart City Index, launched in 2019, uses five key criteria (health and safety, mobility, activities, opportunities, and governance) to assess the economic, people, and technological aspects of smart cities in two principal areas: existing infrastructure, and the technologies and services available to residents.[5] While the motivation for smart city development is evident, there

[1] Bibri, S. E. (2019). On the sustainability of smart and smarter cities in the era of big data: an interdisciplinary and transdisciplinary literature review. *Journal of Big Data*, *6*(1), 1–64; Yigitcanlar, T., Degirmenci, K., Butler, L., & Desouza, K. C. (2022). What are the key factors affecting Smart City transformation readiness? Evidence from Australian cities. *Cities*, *120*, 103434. https://doi.org/10.1016/j.cities.2021.103434.

[2] United Nations. (2023). *The Sustainable Development Goals Report 2023*. https://unstats.un.org/sdgs/report/2023/.

[3] IMD, IMD World Competitiveness Center, SCO Smart City Observatory, SUTD. (2021). *Smart City Index 2021*. https://www.imd.org/smart-city-observatory/home/#_smartCity.

[4] IESE Business School. (2022). *IESE Cities in Motion Index 2022*. https://www.iese.edu/media/research/pdfs/ST-0633-E.pdf.

[5] IMD-SUTD. (2021). *Op. cit.*

is less understanding of the ways to get there.[6] There are knowing-doing gaps on the critical factors that accelerate smart city development, e.g., in the implementation of smart city governance, sustainability transition, and how smart city strategies act out in practice and which processes accompany those developments.[7] This study on 11 smart cities from the SUTD-IMD Index — Auckland, Boston, Copenhagen, Gothenburg, Guangzhou, Hangzhou, Melbourne, Milan, Seoul, Tokyo, and Vancouver — seeks to unpack the critical factors of smart city development using a socio-technical systems approach.[8] It begins with an exploration into the definition of the "what" and "why" smart cities (Chapter 1). In what follows, over 11 chapters, smart city implementation in each of the 11 smart cities is discussed, including the "what" and "how" of people, organisations, and technologies in advancing smart city goals, anchored in their smart city initiative and project case examples.

1.1 Knowing Smart Cities

The concept of smart cities is not new. Its origins can be traced to the concept of "cybernetically planned cities" in the 1970s, the "networked or wired cities" in the 1980s, and the "smart growth movement" in the 1990s.[9] Despite worldwide development, there is no single definition but a multitude of what a smart city is, reflecting in part the nature of smart

---

[6] Lim, C., Kim, K. J., & Maglio, P. P. (2018). Smart cities with big data: Reference models, challenges, and considerations. *Cities, 82*, 86–99.

[7] Ruhlandt, R. (2018). The governance of Smart Cities: A systematic literature review. *Cities, 81*, 1–23.

[8] Geels, F. W. & Schot, J. (2010). The dynamics of transitions: a socio-technical perspective. In J. Grin, J, Rotmans, & J. Schot (Eds.), *Transitions to Sustainable Development: New Directions in the Study of Long Term Transformative Change* (pp. 11–104). Routledge.

[9] Bibri, S. E. (2019). *Big Data Science and Analytics for Smart Sustainable Urbanism: Unprecedented Paradigmatic Shifts and Practical Advancements*. Springer; Grossi, G., & Trunova, O. (2021). Are UN SDGs useful for capturing multiple values of smart city? *Cities, 114*, 103193. https://doi.org/10.1016/j.cities.2021.103193; Nikki Han, M. J. & Kim, M. J. (2021). A critical review of the smart city in relation to citizen adoption towards sustainable smart living. *Habitat International, 108*, 102312. https://doi.org/10.1016/j.habitatint.2021.102312.

city development and in part contextual conditions, needs, and motivations. Fundamentally, the concept of a smart city is inherently interdisciplinary, involving perspectives from urban planning, information technology, engineering, sustainability, and citizen engagement, among others,[10] and necessitating consideration of an array of perspectives and factors in implementation.

The term "smart" is frequently used to describe more innovative, often technology-led solutions to achieve improved urban development. According to the 2021 IMD-SUTD Smart City Index, a smart city is "an urban setting that applies technology to enhance the benefits and diminish the shortcomings of urbanisation for its citizens".[11] The World Smart City Organisation alludes to smart city development "when investments in human, social capital, traditional (transport), as well as modern ICT infrastructure, promote sustainable economic development and high quality of life, with a wise management of natural resources, through participatory actions and commitment".[12] The definitions of smart city often canvass the application of advanced technologies, particularly Information and Communication Technologies (ICT) and Internet of Things (IoT) solutions for more efficient city management. At times, they are ICT-related city descriptions, such as digital city,[13] intelligent city,[14] and knowledge city.[15] The power of smart cities lies in their potential to improve the lives of their populations through the application of advanced technologies and data-driven decisions and operations so that urban functions and services can be optimised and become more efficient, innovative, equitable, connected, secure, and sustainable.

---

[10] Shamsuzzoha, A., Nieminen, J., Piya, S., & Rutledge, K. (2021). Smart city for sustainable environment: A comparison of participatory strategies from Helsinki, Singapore and London. *Cities*, *114*, 103194. https://doi.org/10.1016/j.cities.2021.103194.

[11] IMD-SUTD. (2021). *Op. cit.*, p. 133.

[12] Caragliu, A., Del Bo, C., & Nijkamp, P. (2011). Smart Cities in Europe. *Journal of Urban Technology, 18*(2), 65–82.

[13] Ishida, T. (2002). Digital city Kyoto. *Communications of the ACM*, *45*(7), 76–81.

[14] Komninos, N. (2006). The architecture of intelligent cities: Integrating human, collective and artificial intelligence to enhance knowledge and innovation. In *2006 2nd IET International Conference on Intelligent Environments — IE 06*, Vol. 1, pp. 13–20.

[15] Edvinsson, L. (2006). Aspects on the city as a knowledge tool. *Journal of Knowledge Management*, *10*(5), 6–13.

Increasingly, smart cities are associated with not only technology, data, and connectivity but also people and government involvement. For example, promoting urban resilience in the face of climate and environmental hazards, including extreme weather events and rising sea levels, often mandates comprehensive preparedness strategies, robust disaster recovery plans, community engagement initiatives, and resilient infrastructure capable of withstanding unforeseen disturbances, among others.[16] Fostering economic prosperity in the face of resource scarcity, accommodating population growth, and ensuring resource sustainability also require a nuanced balance of urban, economic, social, and environmental development, which may entail urban planning and policy that champion eco-friendly initiatives, such as green infrastructure and financing, knowledge-based industries, mixed land use policies, and ecologically sensitive development.[17] Overcoming multi-faceted urban challenges transcends beyond digital technology-driven projects to collaborative and interdisciplinary efforts, innovative policies, and a steadfast commitment to creating sustainable, resilient, and inclusive urban environments for all.[18]

Besides technological applications, another basic tent of smart city development is the soft infrastructure of the city, such as relating to effective policies and its "smartness footprint", variously measured with smart city capacity indexes.[19] Another is a reference to the city's innovation capacity, to its evolvement from, e.g., experimentations and living labs for innovation testing[20] to innovative solutions, which are mainly based, but

[16] Asadzadeh, A., Fekete, A., Khazai, B., Moghadas, M., Zebardast, E., Basirat, M., & Kötter, T. (2023). Capacitating urban governance and planning systems to drive transformative resilience. *Sustainable Cities and Society*, *96*, 104637. https://doi.org/10.1016/j.scs.2023.104637.

[17] Bedi, C., Kansal, A., & Mukheibir, P. (2023). A conceptual framework for the assessment of and the transition to liveable, sustainable and equitable cities. *Environmental Science & Policy*, *140*, 134–145.

[18] Bedi *et al.* (2023). *Op. cit.*; Asadzadeh *et al.* (2023). *Op. cit.*

[19] Giffinger, R., Fertner, C., Kramar, H., Kalasek, R., Pichler-Milanović, N., & Meijers, E. (2007). *Smart Cities Ranking of European Medium-sized Cities.* Centre of Regional Science, Vienna University of Technology. https://www.smart-cities.eu/download/smart_cities_final_report.pdf; IMD-SUTD. (2021). *Op. cit.*

[20] Komninos, N. (2009). Intelligent cities: Towards interactive and global innovation environments. *International Journal of Innovation and Regional Development*, *1*(4), 337–355.

not limited to ICT, to solve various urban problems, e.g., to improve liveability and achieve sustainability goals.[21] Against the varying conditions and needs of cities, it is imperative that smart city development takes a more holistic approach, where smart city practices embrace not just technological deployment but also effective policies and people/community involvement while responding to local context and conditions.

Prior literature has shown that smart city development involves a range of dimensions and components, varying according to context, technology use, citizen needs, and governance.[22] Its implementation often entails the assemblage of critical urban systems and infrastructure, both ICT and non-ICT, and requires the engagement of multiple actors and sectors at many political levels. Khatoun and Zeadally (2016) have described the smart city as a complex "system of systems".[23] Others have argued that in smart cities, social-ecological-technical systems support transformative changes at various levels.[24] Das and colleagues (2019) have pinpointed eight smart city components: smart infrastructure, smart environment, smart services, smart governance, smart people, smart living, smart transportation, and smart economy.[25]

While smart cities offer promising solutions, there are potential pitfalls in both the technological and non-technological aspects.[26] Smart cities concentrating on digital solutions, services, and infrastructures (e.g., the priority to attract high-technology industries as growth engines for

---

[21] Reddick, C. & Anthopoulos, L. (Eds.). (2015). *Information and Communication Technologies in Public Administration: Innovations from Developed Countries*. Routledge.

[22] Silva, B. N., Khan, M., & Han, K. (2018). Towards sustainable smart cities: A review of trends, architectures, components, and open challenges in smart cities. *Sustainable Cities & Society, 38*, 697–713.

[23] Khatoun, R. & Zeadally, S. (2016). Smart cities: Concepts, architectures, research opportunities. *Communications of the ACM, 59*(8), 46–57.

[24] Caragliu, A. & Del Bo, C. F. (2016). Do smart cities invest in smarter policies? Learning from the past, planning for the future. *Social Science Computer Review, 34*(6), 657–672; Geels & Schot (2010). *Op. cit.*

[25] Das, A., Mishra Sharma, S., & Ratha, B. (2019). The new era of smart cities, from the perspective of the internet of things. In D. B. Rawat & K. Z. Ghafoor (Eds.), *Smart Cities Cybersecurity and Privacy* (pp. 1–9). Elsevier.

[26] Lam, P. T. I. & Ma, R. (2019). Potential pitfalls in the development of smart cities and mitigation measures: An exploratory study. *Cities, 91*, 146–156; Silva *et al.* (2018). *Op. cit.*

urban economies, the tendency to reframe urban problems into technological problems) without social (people) consideration can lead to technological determinism and raise public risks, including social inequalities and digital divide, eroding the benefits of smart city initiatives.[27] There is a body of work on how technology-driven approaches may prioritise solutions that privilege regions and populations with more access to digital and socio-economic resources, neglecting the needs and preferences of low-income communities and deepening existing social inequalities.[28] Others have highlighted the potential for personal information misuse and data insecurity in smart city initiatives, where reliance on sensors and cameras to collect real-time data for better decision-making can lead to invasive levels of surveillance by both government and private sectors, if not implemented with people's consideration.[29] The suggestion is that smart cities (and their implementation) require attention to multi-dimensional processes, connecting the physical, social, and innovation capital with governance and technology to transition towards more sustainable outcomes.[30]

1.2 Implementing Smart Cities

Cities in both developed and developing countries have pursued smart city development, formulating innovative solutions to various economic, social, and environmental challenges to enhance urban liveability and

---

[27] León, L. F. A. & Rosen, J. (2020). Technology as Ideology in Urban Governance. *Annals of the American Association of Geographers*, *110*(2), 497–506; Kvasny, L. & Keil, M. (2006). The challenges of redressing the digital divide: A tale of two US cities. *Information Systems Journal*, *16*(1), 23–53; Lim, Y., Edelenbos, J. & Gianoli, A. (2019). Identifying the results of smart city development: Findings from systematic literature review. *Cities*, *95*, 102397. https://doi.org/10.1016/j.cities.2019.102397.

[28] Heeks, R. (2022). Digital inequality beyond the digital divide: Conceptualizing adverse digital incorporation in the global South. *Information Technology for Development*, *28*(4), 688–704.

[29] Braun, T., Fung, B. C. M., Iqbal, F., & Shah, B. (2018). Security and privacy challenges in smart cities. *Sustainable Cities and Society*, *39*, 499–507.

[30] Hodson, M. & Marvin, S. (2010). Can cities shape socio-technical transitions and how would we know if they were? *Research Policy*, *39*, 477–485; Rohracher, H. & Spath, P. (2013) The interplay of energy policy and socio-technical transitions: The eco-cities of Graz and Freiburg in retrospect. *Urban Studies*, *51*, 1415–1431.

sustainability. Building a smart city requires consideration, coordination, collaboration, and balancing of many elements in the city, including the following[31]:

- **Technology**: For example, within a smart city ecosystem, IoT is often used to establish intelligent infrastructure, interconnecting physical devices and sensors for real-time monitoring, data collection, and analysis, while facilitating the integration and management of various urban systems and services, such as transportation, energy, waste management, for resource optimisation, efficiency, citizen experiences, and to enable sustainable development within a smart city.[32]
- **Community needs and partnerships**: It is not sufficient to simply adopt technologies in silos. It is important to understand community needs and leverage partnerships and collaborations to combine innovations with people-centric operating strategies so as to realise the full benefits towards more sustainable outcomes and optimised user experiences, e.g., social equity remains a paramount concern in smart city development, requiring a holistic approach that addresses socio-economic disparities, promotes affordable housing initiatives, and enforces equitable access to essential services, especially for disadvantaged populations.[33]
- **Policymaking and governance**: For example, the rapid development of smart city technology has often surpassed existing legal and regulatory frameworks, leading to implementation uncertainty, requiring the review and formulation of clear guidance for stakeholders, such as relating to data privacy, security, and ownership, while reassuring them that smart city innovation and development will not be hindered.
- **Financing**: Allocating funding to smart city development is a critical factor to the speed, scale, and future outcome of any project

---

[31] Yuen, B., Chan, F., Yang, X., & Lim, K. (2024). *Smart Sustainability Transformation Playbook.* World Scientific Publishing Company Incorporated.

[32] Caetano, T., Winkler, H., & Depledge, J. (2020). Towards zero carbon and zero poverty: Integrating national climate change mitigation and sustainable development goals. *Climate Policy*, *20*(7), 773–778.

[33] Meerow, S., Pajouhesh, P., & Miller, T. R. (2019). Social equity in urban resilience planning. *Local Environment*, *24*(9), 793–808.

or initiative within a smart city ecosystem, e.g., funding smart city infrastructure and technology often requires heavy upfront investment as well as long-term financial sustainability, including equipping municipal employees with the training and knowledge of how to use new technology, systems, and data that will be collected to support these smart city initiatives.

Although smart city development is sometimes fraught with challenges and obstacles, and it is not always easy to maintain the right balance of knowledge and execution, the development experience of other smart cities lends a reference point for good practices and potential pitfalls during implementation. The 11 smart city developments reviewed in this book span various geographical contexts and smart city progression (Table 1.1). The cities encompass a range of smart city development on selected critical issues in economic prosperity (future digital economy,

Table 1.1. 11 case study smart cities.

City	IMD-SUTD 2021 Smart City Ranking*	IMD-SUTD 2023 Smart City Ranking†
Economic prosperity		
Auckland	16, AA	22, A
Gothenburg	46, BBB	36, A
Hangzhou	63, CCC	70, CCC
Social inclusion		
Boston	22, BBB	34, A
Guangzhou	66, CCC	71, CCC
Seoul	18, AA	16, AA
Tokyo	67, BB	72, BB
Vancouver	28, A	42, A
Environmental sustainability		
Copenhagen	5, AAA	4, AA
Melbourne	36, BBB	31, A
Milan	69, CC	82, BB

Notes: *,†from IMD Smart City Index 2023; smart city rankings are expressed in terms of position among a list of 141 smart cities worldwide reviewed along 2 dimensions, in terms of numerical and alphabetical ratings. Alphabetical ratings are calculated from the city's performance relative to other cities within the group from AAA to D.

smart retail, and smart tourism), social inclusion (digital inclusion, digital placemaking, smart health service, and smart youth empowerment), and environmental sustainability (climate resilient neighbourhood, circular economy, and smart climate action).

Although discussed separately under three key domains of sustainable development (economic, social, and environmental), it is noted that the smart city initiative and project may often address not one but two or all three areas, demonstrating the interconnectedness between the domains of sustainability. The three thematic domains of sustainable urban development do not exist alone but align with the 17 United Nations Sustainable Development Goals and the 2030 Agenda for Sustainable Development:

- **Economic Prosperity**: For example, employing automation and adapting problem-solving methodologies to mitigate the strain on physical, administrative services, and infrastructure, and encompassing transaction logistics and online marketing to increase efficiency and economic output (e.g., e-commerce, e-wallets, automated logistics, and hybrid stores).[34]
- **Social Inclusion**: For example, promoting social equality involves not only the facilitation of affordable housing, accessible transportation, inclusive public spaces, and racial integration but also other targeted social issues. Social inclusion is also realised through disseminating knowledge, communication, and cultivating social capital through technological means (e.g., online platforms, social media, and digital placemaking).[35]
- **Environmental Sustainability**: For cxample, improvements in transportation systems, reductions in energy consumption, and advancements in the safeguarding and administration of environmental resources and infrastructure (e.g., conservation of water, energy, food supplies, and waste, and mitigation of harmful emissions).[36]

[34] Csukás, M. S. & Szabó, R. Z. (2021). The many faces of the smart city: Differing value propositions in the activity portfolios of nine cities. *Cities*, *112*, 103116. https://doi.org/10.1016/j.cities.2021.103116.

[35] Jabareen, Y. R. (2006). Sustainable Urban Forms: Their Typologies, Models, and Concepts. *Journal of Planning Education and Research, 26*(1), 38–52.

[36] Csukás, M. S. & Szabó, R. Z. (2021). *Op. cit.*; Chatfield, A. T., & Reddick, C. G. (2016). Smart city implementation through shared vision of social innovation for environmental

1.3 The Approach

Against the intricate interplay of people, socio-economic conditions, and complex urban systems, a growing body of innovation studies refers to the relevance of embracing a more holistic understanding of smart city building, focusing on not only a technology-oriented approach but also a multi-dimensional systems perspective, involving, e.g., a context of interconnected urban socio-technical networks and institutions.[37] Smart city development, when reconceptualised through a socio-technical systems lens, mandates more balanced attention on technological innovation, human capital investment, environmental friendliness, and local context comprehension. This book addresses the imperative of taking a holistic approach to smart city development, highlighting the implementation of smart cities that are smart and also socially ecologically responsive and people-centric. The focus is on cities' challenges and courses of action in and around the socio-technical systems and processes to enhance the sustainability and liveability of urban areas. The rest of the chapters in this book explore the emergent issues, enabling technologies, approaches, policies, and case studies of building smart cities.

A case study approach is adopted to distil and demonstrate the challenges, contextual factors, solutions, strategies, and impacts of smart city initiatives and implementations. Up to two project case studies from each city are illustrated; these are chosen based on their smart city potential and the availability of information. The analysis draws on several sources, including literature reviews, secondary data, and interviews with key stakeholders (n = 55).[38] Case studies of smart projects in 11 cities are

---

sustainability: A case study of Kitakyushu, Japan. *Social Science Computer Review*, *34*(6), 757–773.

[37] Branny, A., Møller, M. S., Korpilo, S., McPhearson, T., Gulsrud, N., Olafsson, A. S., Raymond, C. M., & Andersson, E. (2022). Smarter greener cities through a social-ecological-technological systems approach. *Current Opinion in Environmental Sustainability*, *55*, 101168. https://doi.org/10.1016/j.cosust.2022.101168; Nochta, T., Wan, L., Schooling, J. M., & Parlikad, A. K. (2021). A socio-technical perspective on urban analytics: The case of city-scale digital twins. *Journal of Urban Technology*, *28*(1–2), 263–287.

[38] Interviews were conducted in person, over zoom and email communication in 9 of the 11 study cities in 2022. Some cities declined/did not respond to our request for information

identified to understand how people, organisations, and technologies interact to achieve a smart city vision for sustainability. The analysis acknowledges that smart city development involves the engagement of multiple actors and takes place in a social context that, to some degree, will influence the adoption and effectiveness of technologies and ultimately, determine whether they meet end-user satisfaction. Smart city development is pivoted on technological changes, connectivity, and data, but also on people and government involvement and the transformation of urban life practices and conditions. Cities can have one, and more often, several smart city innovations and transition pathways, depending on their smart city vision, policy, plans, and development. In the absence of a one-size-fits-all solution, the case study analysis offers some examples of the different choices that can influence transition towards sustainable urban development through economic prosperity, social inclusion, and environmental sustainability issues.

1.3.1 *Economic Prosperity*

Auckland: Smart Economy (Chapter 2): Smart economic development in Auckland has been guided by a long-term strategic plan, Plan Auckland 2050 since 2012. This plan aims to address three key challenges in Auckland, including high population growth, shared prosperity, and environmental degradation. Within the domain of smart economic development, Auckland has implemented digital economy initiatives in both the private and public sectors to create an innovative and competitive tech ecosystem. Innovation hubs are built and scaled up to encourage agglomerations of technology industries across the Auckland region through the GridAKL innovation programme.[39] At the same time, the public sector is digitised through the provision of online public services and products supported by upgrades in government ICT system efficiency.[40]

and interview. Due to COVID-19 pandemic, we were not able to travel to all study cities.

[39] Auckland Unlimited. (n.d.). *Helping Auckland Business*. https://aucklandunlimited.com/helping-auckland-businesses.

[40] Auckland Technology and Governance Team. (2023 July 1). Interview with Auckland Technology and Governance Team [Personal communication].

Despite these achievements, Auckland still faces several challenges to fully realise its goals of developing an agile, innovative, and inclusive economy. On governance, the right operators of innovative hubs need to be thoughtfully selected to ensure that the hubs provide services, tools, and physical spaces that can appropriately address the needs of the innovation ecosystem. With every transformation, funding is essential, as is realising the technological potential to drive coordinated change through the various government agencies. This is especially the case when transforming the public sector and where funded and coordinated efforts are essential. Equally important is an open culture to experimentation and change. Crucially, smart economic development initiatives also need to be more inclusive, delivering a more significant social impact for the broader communities, especially for urban areas with socio-economic disadvantages.

Gothenburg: Sustainable Tourism (Chapter 3): Digital smart tourism in Gothenburg can be traced to 2018 when the Business Plan for Destination Gothenburg 2018–2020 was published, outlining Gothenburg's goal of doubling its tourism numbers (to 9 million guest nights) by 2030 in a sustainable manner.[41] Tourism is an important growth sector, contributing 10.9% of Gothenburg's total economic revenue in 2019.[42] Sustainable tourism in Gothenburg is technology-enabled; 97% of Gothenburg's public transportation is powered by renewable energy, while 95% of its hotels have eco-certification.[43] Meeting venues, accommodation options, dining establishments, and cultural attractions are located near each other.[44] Local

---

[41] Goteborg & Co. (2018). *Way to Grow*. https://goteborgco.se/uploads/2018/10/Affarsplan_Destination_samt_Goteborg__Co__2018-2020_ENG_lowres_spreads.pdf.

[42] Göteborg & Co. (2019). *Annual and Sustainability Report 2019*. https://goteborgco.se/uploads/2020/03/ENG_GbgCo_Arsredovisning_Hallbarhetsredovisning_2019.pdf.

[43] Global Destination Sustainability. (2022). *Raising the Bar 2022 GDS Index Results Report*. https://www.gds.earth/wp-content/uploads/Raising-the-Bar-2022-GDS-Index-Results-Report.pdf; Bulter, A. (26 July 2022). Exploring Gothenburg's Sustainable Side: A Green Weekend Itinerary. *Lonely Planet*. https://www.lonelyplanet.com/articles/sustainable-gothenburg-getaway.

[44] ICTTP 7. (n.d.). *Gothenburg*. https://www.icttp7.se/practical-info/gothenburg/.

innovations have enabled a sharing economic platform, extending sharing projects from local communities to domestic and international tourists. Smarta Kartan (translated to English: Smart Map) is a website that displays sharing projects in Gothenburg where locals and visitors may rent, borrow, share, exchange products, and contribute to services.[45] Min Stad is another interactive map bulletin board (available on both the web and as an application for mobile phones) where users — locals and visitors — can share their stories, needs, and interests within various categories, such as culture, recreation, cycling, and socialising.[46] In the process, digitalisation has changed people's tourism experience with the provision of hybrid experiences and local shared knowledge, and promoted more sustainable consumption. A survey conducted by the Consumer and Citizen Service in 2021[47] with 960 Gothenburg residents showed that 90% are optimistic about reducing their consumption over the next 5 years through sharing platforms, and 70% would donate items they no longer intend to retain. The rest of the surveyed residents have not heard of such sharing services, with some stating low levels of trust, leading to a lack of sharing participation. A further challenge is the budgetary shortfall of city government, a potential hindrance to organisations in their smart imitative growth and expansion.[48]

Hangzhou: Smart Retail (Chapter 4): Hangzhou's well-developed economy and high GDP are key driving factors of the city's success in establishing a bottom-up digitally based economy through smart retail and marketplaces.[49] The New Retail model, originating from Alibaba in

---

[45] Smart Map. (2022). *About the Smart Map*. Smarta Kartan. https://www.smartakartan.se/om-smarta-kartan/.

[46] Klefbom, S. (2021). *Augmented Urban Values: Virtual Gothenburg as a Place for Citizen Dialogue and Shared Lived Experiences*. http://urn.kb.se/resolve?urn=urn:nbn:se:mau:diva-45934.

[47] *Vårt Göteborg*. (21 June 2021). *Göteborgarna positiva till delningsekonomi. Vårt Göteborg*. https://vartgoteborg.se/konsument/goteborgarna-positiva-till-delningsekonomi/.

[48] Persson, J. (16 June 2022). Interview with Collaborative Economics Gothenburg [Personal communication].

[49] 汪秘书长. (13 July 2022). Interview with Hangzhou Secretary General of the Hangzhou New Retail Association [Personal communication].

Hangzhou, combines digital capabilities and physical assets with the use of analytics and retail-as-a-service business to inform consumer patterns and support back-end logistics.[50] Enabling innovations include hybrid retail shops, such as Freshippo with an unstaffed retail model, simultaneously functioning as a neighbourhood consumption and logistics centre for groceries and daily necessities.[51] Another enabling innovation is E-wallet, predominantly smartphone-based platform applications that allow customers to make payments digitally.[52] In combination, they support the Smart Retail ecosystem that automates the retail and payment infrastructure, expanding the retail economic reach to a global scale.[53] However, this is not without challenges, especially when scaling up hybrid physical innovations, including cyber security, and monopoly issues in China. While the Hangzhou government has established factors to determine the success of hybrid retail and how to scale up,[54] it is still financially taxing for local stores to employ this model due to the financial investment needed to set up the automated infrastructure, despite local subsidies.[55] Anti-monopoly regulations and legislations in 2021 and policies, which address data collection practices,[56] online content, and the use of algorithms to target users of the platform economy further suggest

[50] Achim, A.-L. (10 March 2019). How Jack Ma's 'New Retail' Model Puts China Ahead of the Game. *South China Morning Post.* https://www.scmp.com/magazines/style/news-trends/article/2188823/alibaba-gucci-how-jack-mas-new-retail-model-puts-china

[51] Bird, J. (18 November 2018). Alibaba's 'New Retail' Revolution: What Is It, and Is It Genuinely New? *Forbes.* https://www.forbes.com/sites/jonbird1/2018/11/18/alibabas-new-retail-revolution-what-is-it-and-is-it-genuinely-new/; Hou, Y. (2018). Hema — The Pathfinder of Alibaba's New Retail [Investor's Report]. Herma, Alibaba.

[52] Hsu, S. & Li, J. (2019). *China's Fintech Explosion: Disruption, Innovation, and Survival* (pp. 43–65). Columbia University Press.

[53] Lu, L. (2018). Decoding Alipay: Mobile payments, a cashless society and regulatory challenges. *Butterworths Journal of International Banking and Financial Law, 33*(1), 40–43.

[54] 汪秘书长. (13 July 2022). *Op. cit.*

[55] *Ibid.*

[56] Kharpal, A. (20 August 2021). China Passes Major Data Protection Law as Regulatory Scrutiny on Tech Sector Intensifies. *CNBC.* https://www.cnbc.com/2021/08/20/china-passes-key-data-protection-law-as-regulatory-scrutiny-increases.html.

that e-commerce platforms would have to re-evaluate platform data mining practices.[57]

1.3.2 *Social Inclusion*

Boston: Digital Placemaking (Chapter 5): Placemaking takes a people-centric approach to collectively shape public spaces — mainly vacant and underutilised spaces — into vibrant social, cultural, and civic places that people would like to live, work, and play in.[58] Digital placemaking strategies can include interactive public artwork and events or practical solutions to everyday problems like enhancing accessibility and wayfinding through tech-based means.[59] In 2016, the Boston Transportation Department launched the Guide to Placemaking for Mobility, which adopts a policy framework on both traditional and digital placemaking strategies and community-driven activities to reclaim streets for people.[60] Digital placemaking has since been incorporated into civic projects, such as in the Mayor's Office of New Urban Mechanics (MONUM) projects across housing and education, placemaking, digital trust building, financial planning, city maintenance, road safety, food security, and more.[61] Another is the Plan for Accessible Transit Infrastructure (PATI) programme, wherein a digital signage initiative was started in 2019 to provide all passengers with the most accurate real-time transit information about public transportation services.[62] Differently-abled passengers are supported by

[57] Kharpal, A. (30 August 2021). China's Regulatory Crackdown Has Wiped Billions Off Tech Stocks — Here are the Risks Ahead. *CNBC*. https://www.cnbc.com/2021/08/30/china-tech-crackdown-experts-warn-on-the-risks-ahead-for-stocks.html.

[58] A Better City. (2016). *A Guide to Placemaking for Mobility*. https://www.abettercity.org/docs-new/Guide_to_Placemaking_for_Mobility.pdf; Loh, C. G. (2019). Placemaking and implementation: Revisiting the performance principle. *Land Use Policy*, *81*, 68–75.

[59] Halegoua, G. & Polson, E. (2021). *Op. cit.*; Stokes *et al.* (2021). *Op. cit.*; Pang *et al.* (2020). *Op. cit.*

[60] A Better City. (2016). *Op. cit.*

[61] City of Boston. (n.d.). *New Urban Mechanics*. https://www.boston.gov/departments/new-urban-mechanics.

[62] Massachusetts Bay Transportation Agency (MBTA). (n.d.). *Digital Signage*. https://www.mbta.com/projects/digital-signage?from=search-header-desktop&query=digital%20si.

including audio-equivalent resources for riders who are sight-impaired, using large fonts to ensure readability while considering viewing distance, lighting, and other environmental factors.[63] Challenges vary across different placemaking projects due to site and community-specific conditions. Key provisions to enable such innovations include access to infrastructure like 24-hour power and maintenance support. Consistent workforce, inter-departmental engagement, and community support are other critical factors that promote the successful scaling of innovations.[64] Post-implementation evaluation is also useful, revealing ways to make digital signages more effective in information provision and on best placements of signages.[65]

Guangzhou: Digital Inclusion (Chapter 6): Digital inclusion plays a prominent role in Guangzhou's smart city vision by making services more efficient and accessible to citizens through upskilling and expanded use of e-platforms.[66] Digital inclusion also encompasses services catered towards minority groups with plans to utilise ICT in systematic ways for people with disabilities, and in rehabilitation, education, employment, poverty alleviation, social support, and social security.[67] "Suihaoban" (穗好办) consolidates all municipal services into an online application to provide convenient access to all residents, removing the need to physically visit municipal offices. Automated data reconciliation through this service plays a major role in tackling illegal activities, such as false declarations and false documentation.[68] Guangzhou's Regional Health Information

[63] *Ibid.*

[64] Deshpande, Y. (6 July 2022). Interview with the Boston Mayor's Office of New Urban Mechanics [Personal communication].

[65] Subramanian, K. (13 July 2022). Interview with the MBTA Customer Technology Department [Personal communication].

[66] Guangzhou Government. (2020). Guangzhou International. http://www.gz.gov.cn/guangzhouinternational/aboutgz/content/post_5734870.html.

[67] Lin, Z., Yang, L., & Zhang, Z. (2018). To include, or not to include, that is the question: Disability digital inclusion and exclusion in China. *New Media & Society, 20*(12), 4436–4452.

[68] Wang, J. (12 November 2018). Tech Giant Blazes Trail for Profound Digital Change. *Chinadaily.com.cn*. https://www.chinadaily.com.cn/cndy/2018-11/12/content_37242235.htm.

Platform is another innovation, which provides consolidated online healthcare records and services so residents can visit any regional healthcare outlets or have access to telemedicine with transferrable medical records.[69] Despite the progress in inclusive digital services, efforts to provide equal access to social and healthcare support remain urgent, especially for populations who are illiterate, ageing, and without access to the internet.[70] Recommendations include policies designed to close digital divide, including helping people with the use of technology and accessing electronic hospital and technical guidance for older adults through training sessions.[71]

Seoul: Smart Youth Empowerment (Chapter 7): South Korea, including its capital city of Seoul, has a comparatively high prevalence of internet addiction among children and adolescents. Prolonged internet use among adolescents can cause social isolation and mental health problems, such as depression, change in character like aggression, and unhealthy behaviours, such as the lack of exercise and low academic performance.[72] The government has initiated internet addiction policies, introduced internet prevention and counselling programmes, and conducted research studies.[73] The I Will Centre in Seoul is an initiative that contributes to digital youth empowerment by providing digital education and training

[69] Zhang, L., Wang, H., Li, Q., Zhao, M.-H., & Zhan, Q.-M. (2018). Big data and medical research in China. *BMJ*, *360*, j5910. https://doi.org/10.1136/bmj.j5910.

[70] Magsamen-Conrad, K., Wang, F., Tetteh, D., & Lee, Y.-I. (2020). Using technology adoption theory and a lifespan approach to develop a theoretical framework for ehealth literacy: Extending UTAUT. *Health Communication*, *35*(12), 1435–1446.

[71] Tu, J., Shen, M., Zhong, J., Yuan, G., & Chen, M. (2021). The perceptions and experiences of mobile health technology by older people in Guangzhou, China: A Qualitative Study. *Frontiers in Public Health*, *9*, 683712. https://doi.org/10.3389/fpubh.2021.683712.

[72] Yi, Yunjeong, Hyun, Sook-Jung, Lee, Jinhwa, & An, Ji-Yeon. (2017). The Impact of Internet Addiction on Depression among Korean Adolescent Middle and High School Students. *International Journal of Contents*, *13*(2), 6–13.

[73] Cho, E. (2017). Therapeutic interventions for treatment of adolescent internet addiction — Experiences from South Korea. In C. Montag & M. Reuter (Eds.) *Internet Addiction. Studies in Neuroscience, Psychology and Behavioral Economics.* Cham: Springer.

opportunities to combat internet addiction.[74] The Jump Up Internet Rescue boot camp is another initiative to address internet addiction by providing outdoor recreation activities alongside practical cybersecurity and digital privacy education for young people.[75] Based on the individual's physical and mental development stage, educational programmes are accordingly tailored and adapted. Similar educational programmes are being made for parents and teachers as well. By providing young people with the skills, knowledge, and experience they need to succeed in a digital society, these initiatives help create a more resilient and adaptable workforce while promoting social and economic development in the broader community. Despite the government's early detection of the social issue, both Seoul and South Korea face various challenges, including the accelerated pace of internet development. Individuals who are addicted to the internet tend to deny their addiction, and they do not realise that it is a problem.[76] Another challenge is the lack of research and knowledge on the topic of internet addiction.[77] Therefore, while such programmes are a good step towards digital youth empowerment, they need to remain adaptable to the changing online ecosystem and its risks.

Tokyo: Smart Health Services (Chapter 8): Against Japan's fast ageing society,[78] the country has embraced healthcare information technology as

---

[74] Seoul Solution. (2014). Youth internet addiction prevention programs. https://www.seoulsolution.kr/en/content/2073.

[75] Lee, S.-Y., Kim, M. S., & Lee, H. K. (2019). Prevention strategies and interventions for internet use disorders due to addictive behaviors based on an integrative conceptual model. *Current Addiction Reports*, *6*(3), 303–312.

[76] Best, P., Manktelow, R., & Taylor, B. (2014). Online communication, social media and adolescent wellbeing: A systematic narrative review. *Children and Youth Services Review, 41*, 27–36.

[77] Jeong, S. (21 October 2019). The Teenagers so Addicted to Cellphones They're Going to Detox Centers. *CNN*. https://edition.cnn.com/2019/10/20/asia/smartphone-addiction-camp-intl-hnk-scli/index.html.

[78] *Kyodo News*. (2018). Japan's Population Decline to 127 mil., Aging Continues in 2017. https://english.kyodonews.net/news/2018/04/1fc13074ea8a-japans-population-decline-aging-continue-in-2017.html.

a vehicle for reform since the early 2010s.[79] In 2014, the Ministry of Economy, Trade, and Industry (METI) and the Ministry of Health, Labour, and Welfare (MHLW) introduced the notion of next-generation robots as device assistants in both healthcare and everyday life. The technology deployment aims to address the challenges of labour shortages in the healthcare sector.[80] One notable innovation is the Paro robot, which is shaped like a baby harp seal, weighs the same as a newborn child, and is designed to replicate the psychological, physiological, and social benefits of animal therapy.[81] Considering the projected increase in the number of older adults living alone, other robots such as PaPeRo, "Partner-type-Personal-Robot", have been introduced to aid older people to live independently.[82] PaPeRo uses the latest speech synthesis technology and can speak in a natural or cute voice. With built-in face detection, PaPeRo can also locate people from its image recognition technology while identifying up to 30 faces. Funding and support are essential in innovating and creating new robotic technology. The national R&D programme for robotics is planned and funded by the Economic Revitalization Policy and Science, Technology, and Innovation Policy,[83] which significantly lowers the overall cost of these robots for local healthcare uses. However, the cost of individual robots remains too high for the general population, acting as

---

[79] Abraham, C., Nishihara, E., & Akiyama, M. (2011). Transforming healthcare with information technology in Japan: A review of policy, people, and progress. *International Journal of Medical Informatics*, *80*(3), 157–170.

[80] Ministry of Economy, Trade, and Industry. (2014). *Japan's Robot Strategy was Compiled*. https://www.meti.go.jp/english/press/2015/0123_01.html.

[81] Inoue, K., Wada, K., & Shibata, T. (2021). Exploring the applicability of the robotic seal PARO to support caring for older persons with dementia within the home context. *Palliative Care and Social Practice*, *15*, 26323524211030284.

[82] Phys Org. (2013). *NEC Introduces the PaPeRo Petit Robot*. https://phys.org/news/2013-11-nec-papero-petit-robot.html.

[83] International Federation of Robotics. (2021). *World Robotics R&D Programs*. https://ifr.org/downloads/papers/Executive_Summary_-_World_Robotics_RD_Programs_V02.pdf.

the most significant obstacle due to the lack of economic feasibility in deploying these robots.[84]

Vancouver: Smart Governance (Chapter 9): Vancouver's Digital Strategy was prepared by the City of Vancouver Council to enhance multidirectional digital connections among citizens, employees, businesses, and government.[85] A 5-to-7-year plan was imagined and developed in four parts, consulting and implementing ideas from industry insiders, digital thought leaders, start-up founders, business executives, citizens, and city staff.[86] VanConnect, introduced in May 2015, is a mobile app developed by the city to provide an e-city service ecosystem. The main goal is to ensure citizens and businesses can easily interact with the City Council and its services through digital channels while providing access to technology and education to increase digital literacy.[87] Alongside this is the free public Wi-Fi, known as VanWifi, which aims to enhance citizens' daily experience through online, mobile, and social media channels.[88] Developed over three phases, the project involves working with local telecom partners to expand coverage across various parts of the city. As of 2020, there are 521 free public Wi-Fi locations across Vancouver, making VanWiFi one of the largest free public Wi-Fi networks in North America.[89] Overall, the implementation of smart governance strategies has progressed the city from taking an *ad hoc* and siloed approach to a more integrated and strategic approach that brings value to citizens, businesses, and large. However, post-implementation evaluation for the effectiveness and uptake of these initiatives is needed.

---

[84] Inagaki, K. (8 October 2014). Japan Aims to Turn Robotics into Profit. *CNBC.* https://www.cnbc.com/2014/10/08/japan-aims-to-turn-robotics-into-profit.html.

[85] City of Vancouver. (2013). *City of Vancouver Digital Strategy.* https://vancouver.ca/files/cov/City_of_Vancouver_Digital_Strategy.pdf.

[86] City of Vancouver. (n.d.a). *Timeline of Digital Strategy Achievements.* https://vancouver.ca/your-government/digital-strategy-achievements.aspx.

[87] *Ibid.*

[88] City of Vancouver. (n.d.b). *Featured Project: #VanWiFi.* https://vancouver.ca/your-government/vanwifi.aspx.

[89] *Ibid.*

1.3.3 *Environmental Sustainability*

Copenhagen: Climate Resilience Building (Chapter 10): The Copenhagen 2025 Climate Plan is Copenhagen's ambitious strategy to become the world's first carbon-neutral capital by 2025. But in 2022, Copenhagen gave up its net-zero commitment due to funding, generating critical examination on net-zero projects and efforts among global cities. Meanwhile, Copenhagen has embarked on an aggressive climate adaptation strategy, including forming Denmark's first climate-resilient neighbourhood, St Kjelds. The Climate-Resilient Neighbourhood initiative is part of Copenhagen's overall strategy for district and urban renewal in 2010 through the Copenhagen Climate Adaptation Plan, seeking to safeguard Copenhagen and its population against the risks of future extreme weather events.[90] Green solutions at the street level have been implemented in St. Kjelds to efficiently lead stormwater to the harbour, where it can be released without causing any damage to the city landscape. One key area is Bryggervangen, an existing 2,300-foot-long road with a cloudburst management strategy that lined the road with rain gardens to receive rainwater from sidewalks and adjacent residential buildings while allowing excess rainwater into large infiltration ponds at the square.[91] Another area in St. Kjelds is Tåsinge Square, a green oasis with the capacity to manage large volumes of rainwater while serving as a new meeting place for residents. The 1,000-square-metre urban wilderness landscape can delay flooding and provide drainage of rainwater from the surrounding area of approximately 8,000 square metres.[92] The Climate Resilient Neighbourhood initiative improves urban quality of life, enhancing social cohesion, promoting biodiversity, and demonstrating that

---

[90] City of Copenhagen. (n.d.). *Climate Adaptation*. https://international.kk.dk/about-copenhagen/liveable-green-city/climate-adaptation.

[91] Holmes, D. (12 February 2020). Sankt Kjelds Square and Bryggervangen Wins Arne of the Year Award. *World Landscape Architecture*. https://worldlandscapearchitect.com/sankt-kjelds-square-and-bryggervangen-wins-arne-of-the-year-award/.

[92] Danish Architecture Center. (n.d.). *Tåsinge Plads: A Pioneering Project in Denmark's First Climate Neighborhood*. https://dac.dk/en/knowledgebase/architecture/taasinge-plads-a-pioneering-project-in-denmarks-first-climate-neighborhood/.

resilience-building efforts could contribute to sustainable development by delivering multiple benefits beyond climate adaptation.

Melbourne: Smart Climate Action (Chapter 11): Smart climate action is being achieved through multiple heat island mitigation and urban greening strategies. The 4°C Cooler Programme was launched in 2012 by strategically increasing urban forest and public open spaces across the municipality. The aim is to increase the resilience of urban forests to cool Melbourne's summertime temperatures by 4°C.[93] This strategy was developed in response to the extreme heatwave and severe drought experienced between 1995 and 2009. The Green Our City Strategic Action Plan is a wider private realm initiative that encompasses greening initiatives across various sectors, such as the implementation of green laneways, green roofs, and green walls and facades.[94] Apart from urban greening, sensors are deployed in parks to collect real-time data about microclimates, air quality, pedestrian activity, park activity, rubbish bin levels, bench usage, and tree canopies. An example is Argyle Square in Carlton. Serving as the living lab for a collaborative pilot project since 2020, the city uses sensors and other smart devices to transform the public open space into a 'smart park'.[95] In relation to the 4°C Cooler Programme, the microclimate and tree canopy sensor data are collected and analysed to evaluate the effectiveness of the current approach to cooling the city through green infrastructure and increased tree canopy cover. Implementing open space sensing presents challenges in achieving scalability for efficient sensing across diverse areas, demanding strategic and technological competency. Continuous effort is needed to effectively communicate the purpose and

[93] City of Melbourne. (n.d.a). *Cooling Our City*. https://www.melbourne.vic.gov.au/about-melbourne/sustainability/pages/cooling-our-city.aspx; UN Habitat. (2020). *4°C cooler — Using Green Infrastructure to Build a Climate Resilient and Prosperous Melbourne*. https://www.urbanagendaplatform.org/best-practice/4degc-cooler-using-green-infrastructure-build-climate-resilient-and-prosperous.

[94] City of Melbourne. (2017). *Green Our City Strategic Action Plan 2017–2021: Vertical and Rooftop Greening in Melbourne*. https://www.melbourne.vic.gov.au/community/greening-the-city/green-infrastructure/Pages/green-our-city-action-plan.aspx.

[95] City of Melbourne. (n.d.b). *Data in the Park — Argyle Square*. https://participate.melbourne.vic.gov.au/emerging-tech-testbed/data-park.

value of public space sensing to build trust, clarify data collection processes, and inform citizens. Optimising the value of sensing data within council departments requires overcoming communication barriers by establishing standardised formats and protocols and providing council employees with training for meaningful data utilisation.[96]

Milan: Sustainable Food Availability (Chapter 12): Milan's plans to tackle the significant social and environmental issues caused by urban food waste and transition towards a circular economy through a range of strategies, including greater redistribution of food losses across private and public sectors, business waste tax reduction to incentivise surplus food donations, and greater local and international outreach on food waste.[97] The Milan Food Policy Office was established in 2017 and has since engaged Milan's many assets in the city and rural agricultural regions for circular economy development. It acts on the urban food system around five key priorities: food access, education, food production, food waste management, and scientific research.[98] The City of Milan was the first Italian city to research food issues and consequently implement its Urban Food Policy, disseminated through the Office of the Milan Urban Food Policy Pact (MUFPP). A set of indicators was developed by the MUFPP team, which was used to start impact evaluation on the Milan food system.[99] This provides impetus to local authorities of the metropolitan area to follow the example of Milan in working towards a more sustainable approach for their food systems. As a result of these initiatives, social, health, and environmental benefits have been registered. The impacts and innovations of Milan's food system are shared regionally and with the international community. However, the impacts are largely isolated, and the system can be further improved by scaling up local initiatives and

[96] City of Melbourne. (n.d.c). *Data in the Park — Eades Park.* https://participate.melbourne.vic.gov.au/data-park-eades-park.

[97] Gordon, M. & Bidault, M. (2019). *Cities 100 Report.* C40 Cities, Realdania, Nordic Sustainability.

[98] Porreca, E. (9 June 2022). Interview with Milan Food Policy [Personal communication].

[99] Urban Sustainability Exchange. (n.d.). *The Milan Food Policy.* https://use.metropolis.org/case-studies/the-milan-food-policy.

working with a denser network of partners for a wider impact. At the global level, challenges faced by cities encompass overly ambitious goals, challenging political situations, a lack of capacity, insufficient participation from critical actors, conflicting national policies, and inadequate engagement mechanisms, which are impeding long-term sustainable development.[100] A useful strategy to solve these challenges is through peer-to-peer learning and cross-pollination of international knowledge and good practices, which MUFPP is currently embarking on.

---

[100] *Ibid.*

CHAPTER 2

AUCKLAND: SMART ECONOMY

2.1 City Overview

Location	North Island of New Zealand
Land Size	572.48 km^2 (2018)
Climate	Subtropical
Geography (topology/ landscape)[101]	Known as the "City of Sails", Auckland is located between two fine harbours. Its unique landscape also features 48 volcanic cones
Environmental Risks[102]	High carbon emission in the transportation sector, flooding risks, and ineffective on-site wastewater treatment in some areas
Population	1.72 million (2021)
GDP	US$83.7 billion (2022)
GDP Per Capita[103]	US$42,625 (2021)
Key Urban Issues of Smart City	Skill shortages in tech sector, spatial inequality in tech industry development, and lack of diversity in tech workforce

[101] New Zealand Immigration. (n.d.). *Auckland*. https://www.live-work.immigration.govt.nz/choose-new-zealand/regions-cities/auckland#:~:text=The%20city%20covers%205%2C000%20square,cones%20it%20is%20built%20around.

[102] Auckland Council. (2018). *Environment and Cultural Heritage Explained*. https://www.aucklandcouncil.govt.nz/plans-projects-policies-reports-bylaws/our-plans-strategies/auckland-plan/environment-cultural-heritage/Pages/environment-and-cultural-heritage-explained.aspx.

[103] Stats NZ. (2021). *Regional Gross Domestic Product: Year Ended March 2020*. https://www.stats.govt.nz/information-releases/regional-gross-domestic-product-year-ended-march-2020.

Located between two harbours in the North Island of New Zealand, Auckland is known as the "City of Sails" with the highest recreational boats per capita in the world.[104] The urban area of Auckland region spans over 572.48 km^2 in 2018.[105] Between 1996 and 2018, the Auckland region had the highest overall rate of urban expansion by square kilometres (72.59), followed by Canterbury (57.3) and then Waikato (48.45).[106] Auckland's climate is subtropical, featuring warm humid summers and mild damp winters.[107] The summer months are between December and March (averaging around 24°C), while the winter months are between June and August (averaging around 16°C).[108]

Auckland is the largest city in New Zealand by population.[109] The estimated population of the Auckland region was 1.72 million in 2021, accounting for about 34% of New Zealand's population.[110] Auckland's population is projected to exceed 2.3 million by 2048, accounting for 37% of national population.[111] The city is largely made up of a young community, with a median age of 34.6 years in 2018.[112] It is also the most ethnically

[104] New Zealand Immigration. (n.d.). *Op. cit.*

[105] Ministry for the Environment (MfE). (n.d.). *Land Use Data.* https://data.mfe.govt.nz/tables/category/environmental-reporting/land/land-use/.

[106] Statistics New Zealand. (n.d.). *Urban Land Cover.* https://www.stats.govt.nz/indicators/urban-land-cover/.

[107] Chappell, P. R. (n.d.). *The Climate and Weather of Auckland* (2nd edition). https://niwa.co.nz/static/Auckland%20ClimateWEB.pdf.

[108] University of Auckland. (n.d.). *About Auckland.* https://www.auckland.ac.nz/en/on-campus/life-on-campus/living-in-auckland/about-auckland.html#:~:text=The%20city%20is%20surrounded%20by,and%20the%20adjoining%20Hauraki%20Gulf.

[109] Stats NZ. (n.d.). *2018 Census Place Summaries.* https://www.stats.govt.nz/tools/2018-census-place-summaries/.

[110] Auckland Council. (n.d.). *Auckland's Population.* https://www.aucklandcouncil.govt.nz/plans-projects-policies-reports-bylaws/our-plans-strategies/auckland-plan/about-the-auckland-plan/Pages/aucklands-population.aspx.

[111] *Ibid.*; Auckland Unlimited. (2023). *Population Dashboard.* https://industry.aucklandnz.com/auckland-index/people.

[112] Auckland Council. (2018). *Auckland Plan 2050.* https://www.aucklandcouncil.govt.nz/plans-projects-policies-reports-bylaws/our-plans-strategies/auckland-plan/about-the-auckland-plan/docsprintdocuments/auckland-plan-2050-print-document.pdf.

diverse region, with over 180 ethnicities;[113] 13% of Auckland's population are of Māori descent.[114] Due to the nature and availability of jobs, immigration to New Zealand is heavily concentrated towards Auckland, where 42% of Aucklanders were born overseas.[115] However, with the COVID-19 pandemic and international travel restrictions, Auckland's population growth has declined for the first time in 2021 (annual population growth of –0.6%).[116] The population decline continues in 2022 (–0.5%).[117]

Auckland is the economic powerhouse of New Zealand. In 2022, the GDP of the Auckland region was NZ$137.83 billion (US$83.65 billion), accounting for 37.6% of the national GDP.[118] Auckland's GDP per capita in 2021 (the latest available statistics) was NZ$71,978 (US$42,625).[119] Over the past decade, Auckland had an average economic growth rate of 3.4% per year, which is slightly higher than the national economy of 3.0% per year.[120] The fastest economic growth in Auckland was in 2003 (5.5%), while the lowest was in 2009 (–2.5%) during the global financial crisis.[121] Auckland's economy is driven by several key sectors, including screening, technology, construction, advanced manufacturing, circular economy, food and beverage manufacturers, and international education.[122] Knowledge-intensive workforce has grown substantially over the past two

---

[113] Auckland Unlimited. (2023). *Op. cit.*

[114] *Ibid.*

[115] *Ibid.*

[116] Statistics New Zealand. (2021). *Auckland's Population Falls for the First Time*. https://www.stats.govt.nz/news/aucklands-population-falls-for-the-first-time.

[117] Auckland Unlimited. (2023). *Op. cit.*

[118] Auckland Unlimited. (2023). *Business and Economy Dashboard*. https://industry.aucklandnz.com/auckland-index/business-and-economy.

[119] Stats NZ. (2021). *Regional gross Domestic Product: Year Ended March 2020*. https://www.stats.govt.nz/information-releases/regional-gross-domestic-product-year-ended-march-2020.

[120] Infometrics. (2023). *Auckland Overview*. https://ecoprofile.infometrics.co.nz/auckland/pdfprofile#:~:text=Auckland%20accounted%20for%2037.4%25%20of%20national%20GDP%20in%202022.

[121] *Ibid.*

[122] Auckland Unlimited. (n.d.). *Sectors of Focus*. https://industry.aucklandnz.com/business/economy-and-sectors/sectors-of-focus.

decades, with the sector accounting for almost half of Auckland's employment in 2022.[123] During the COVID-19 pandemic (2020–2022), while Auckland sustained an economic growth of 1.3%, the negative impact was particularly felt in some industries, such as accommodation, food and beverage services, transport, postal, and warehousing, due to lockdowns and international travel restrictions. The sector of professional, scientific, and technical services made the most significant contribution to Auckland's economic growth between 2021 and 2022.[124]

Auckland is governed by the Auckland Council, which is responsible for all government decisions and responsibilities in the Auckland region. The Auckland Council was established in November 2010 to replace seven territorial authorities and the Auckland Regional Council, with the intent to deliver strong regional decision-making while meeting diverse local needs and interests.[125] The governance structure of Auckland Council encompasses the governing body and 21 local boards.[126]

2.2 Smart City Development: Quality Urban Growth and Development

Auckland's vision, established in the Auckland Plan 2050, is to become a place that Aucklanders love and are proud of, a place they want to stay in or return to, and a place that others want to visit, move to, or invest in.[127] The Auckland Plan 2050 is a long-term strategic plan to guide Auckland's development to 2050, addressing three key challenges, namely, high population growth, shared prosperity, and environmental degradation. The plan was first released in 2012 as a landmark document, encompassing a series of objectives and targets for Auckland's economic, social, and

[123] Auckland Unlimited. (2023). *Op. cit.*

[124] *Ibid.*

[125] Auckland Council. (n.d.). *What is Auckland Council.* https://governance.aucklandcouncil.govt.nz/2-overview/what-is-auckland-council/.

[126] *Ibid.*

[127] Auckland Council. (2018). *What Will Auckland Look Like in the Future?* https://www.aucklandcouncil.govt.nz/plans-projects-policies-reports-bylaws/our-plans-strategies/auckland-plan/development-strategy/future-auckland/Pages/what-auckland-look-like-future.aspx.

environmental development. In June 2018, the Auckland Plan 2050 was updated into a more streamlined spatial plan, enabling a simpler structure and clearer links between outcomes, directions, and measures.[128]

2.2.1 *Quality Urban Growth*

The Auckland Plan 2050 identified six key areas for significant change to occur over the next three decades[129]:

- Belonging and participation, where all Aucklanders will be part of and contribute to society, access opportunities, and have the chance to develop to their full potential.
- Māori identity and well-being, where Māori's identity, well-being, and culture are advanced, promoted, and benefitted.
- Homes and places, where it ensures citizens live in secure, healthy, and affordable homes and have access to a range of inclusive public places.
- Transport and access, ensuring Aucklanders can get where they want to go more easily, safely, and sustainably.
- Environment and cultural heritage, where Aucklanders preserve, protect, and care for the natural environment as its shared cultural heritage.
- Opportunity and prosperity, where Auckland is prosperous with many opportunities and delivers a better standard of living for everyone.

For each area, the Auckland Plan outlines the desirable outcomes and their importance for Auckland's future and provides development strategies to achieve those outcomes. In total, the plan contained 20 directions and 37 focus areas. The Development Strategy (an integral element of the Auckland Plan) was developed to guide Auckland's development towards greater sustainability in the economic, social, cultural, and environmental domains.[130]

---

[128] Auckland Council. (2018). *Op. cit.*

[129] *Ibid.*

[130] *Ibid.*

2.2.2 *Smarter Urban Development*

Smarter and more innovative solutions are identified in several of the strategic directions and focus areas. For instance, at the macro-level of urban spatial structure, Auckland takes a "quality compact approach" to growth and development. Compactness will be achieved through focusing future development in and around existing and new urban cores to limit urban sprawl into the rural hinterland. Quality urban growth and development will be supported through enhanced accessibility by public transportation and active mobility, improved walkability to local amenities, more efficient land use, and coordinated delivery of infrastructure.[131]

To advance belonging and participation, the Southern Initiative was established in 2012 as a place-based regeneration programme to drive social and community innovation in South Auckland, a major urban area at risk of worsening intergenerational socio-economic vulnerability.[132] This programme seeks to bring about inclusive investment and cross-sectoral governance actions to deliver transformational social, economic, and physical outcomes in South Auckland. The programme embraces social innovations that are purposefully designed to optimise the use of available resources and empower the local community to fulfil their human and economic development potential. The programme's innovation capacity is further supported by the Co-design Lab, which became an official part of the Southern Initiative in 2017.[133] The Co-design Lab was funded by Auckland Council and sponsored by eight central government departments to provide a neutral space to test the use of co-design and other innovative solutions for complex social issues.[134]

---

[131] Auckland Council. (2018). *How Auckland Will Grow and Change — A Quality Compact Approach.* https://www.aucklandcouncil.govt.nz/plans-projects-policies-reports-bylaws/our-plans-strategies/auckland-plan/development-strategy/future-auckland/Pages/how-auckland-grow-change-quality-compact-approach.aspx.

[132] Auckland Council. (2018). *Op. cit.*; The Southern Initiative. (n.d.). *Our History.* https://www.tsi.nz/our-history.

[133] The Southern Initiative. (n.d.). *Op. cit.*

[134] Auckland Council. (2018). *Op. cit.*

Smart innovation is also embraced in other sectors, such as housing, environment, and transportation. More specifically, the housing sector calls for innovative designs to help accelerate quality housing development at a scale that improves housing choices. For better environment, technological innovation in the construction and building industries is embedded in sustainable design practices to protect and enhance the natural environment.[135] The transportation sector supports a "fail fast" culture when testing and integrating a wide range of emerging transport technologies and services. Big data and open data are utilised to enhance transport network management and inform Aucklanders' travel decisions.[136] In response to climate change, Auckland recognises the numerous benefits of transitioning to a low-carbon economy and prioritises investment in more resilient infrastructure.[137] The following section focuses on smart growth strategies that Auckland has developed and implemented towards greater economic sustainability.

Auckland Smart City Timeline

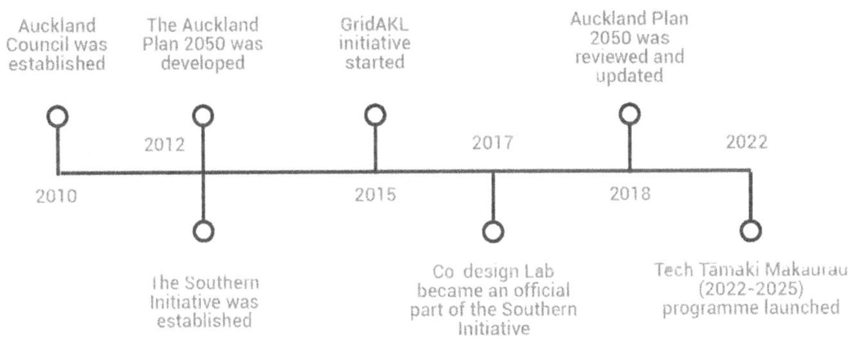

2.3 Smart Growth Strategy

Against the rapidly changing pace of technology and global competitiveness, Auckland's economy seeks to be constantly agile and innovative.[138]

---

[135] *Ibid.*

[136] *Ibid.*

[137] *Ibid.*

[138] *Ibid.*

To guide smart economic growth, a series of coordinated initiatives have been developed and implemented at various governance levels, ranging from the national and regional governments to Auckland's economic agency.

At the national level, the Ministry of Business, Innovation and Employment (MBIE) released the 2015 "Business Growth Agenda", with the vision to build a productive and competitive economy by 2025. This vision is to be achieved through six focus areas: export markets, investment, innovation, skilled and safe workplaces, natural resources, and infrastructure. In the area of building innovation, a Digital Economy Programme was established in late 2015, with the aim of enabling New Zealand to transition to a leading digital nation — "a nation with a thriving digital sector, where our businesses, people, and government are all using digital technology to drive innovation, improve productivity, and enhance the quality of life for all New Zealanders".[139]

More recently, in May 2023, the MBIE released the Digital Technologies Industry Transformation Plan (ITP) 2022–2032 to guide the growth of innovative industries in New Zealand. The vision of the ITP is "The world looks to Aotearoa New Zealand as a leader in ethical, innovative, inclusive and sustainable digital technologies. These technologies enable our economy to prosper, help our businesses to grow stronger and compete internationally, and contribute to the well-being of all New Zealanders".[140] The ITP identifies four immediate focus areas:

- Growing export success with an initial focus on the Software-as-a-Service (SaaS) sub-sector and game development as a future priority sub-sector.

---

[139] Ministry of Business, Innovation and Employment (MBIE). (2015). *Business Growth Agenda — Building a Digital Nation, Part of BGA Building Innovation.* https://www.mbie.govt.nz/dmsdocument/3226-building-a-digital-nation-bga-pdf.

[140] MBIE. (2023). *The Digital Technologies Industry Transformation Plan (ITP).* https://www.mbie.govt.nz/business-and-employment/economic-development/industry-policy/industry-transformation-plans/digital-technologies/.

- Showcasing New Zealand's tech story in terms of its current scope and potential.
- Enhancing the skills and talent pipeline to support the growth of the digital technologies sector.
- Enriching Māori inclusion and enterprise in the digital technologies sector.

In line with the national strategies on digital economy, the Auckland Council sets out three strategic directions for smarter and more inclusive economic growth in the Auckland Plan 2050[141]:

- Creating the conditions for a resilient economy through innovation, employment growth, and enhanced productivity.
- Attracting and retaining skills, talent, and investment.
- Developing skills and talent for the changing nature of work and lifelong achievement.

Under these strategic directions, Auckland focuses on[142]

- Harnessing emerging technologies and ensuring equitable access to high-quality digital data and services.
- Ensuring regulatory planning and other mechanisms to support business, innovation, and productivity growth.
- Advancing Māori employment and supporting Māori business and organisations to be significant drivers of Auckland's economy.
- Leveraging Auckland's position to support growth in exports.
- Increasing educational achievement, lifelong learning, and training, with a focus on those most in need.

Governed by the Auckland Council, Tātaki Auckland Unlimited — Tātaki Makaurau Auckland's economic and cultural agency — is responsible for driving economic growth and industry innovation. The mandate

---

[141] *Ibid.*
[142] *Ibid.*

of Tātaki Auckland Unlimited is "to help Auckland become a major Asia-Pacific innovation hub by building a culture of innovation and entrepreneurship, and growing and retaining skilled talent".[143] More specifically, Tātaki Auckland Unlimited developed the Tech Tāmaki Makaurau, 2022–2025, a 3-year strategic programme to accelerate the growth of Auckland's technology industry and to create jobs and attract talent and investment.[144] Auckland's tech sector consists of Information and Communications Technology (ICT), high-tech manufacturing, and biotech businesses.[145] As New Zealand's innovation hub, Auckland's technology industry is dynamic and growing. Auckland is home to 54% of the nation's tech sector and 120 of Aotearoa's top 209 tech companies. The tech sector contributes NZ$13.6 billion to Auckland region's GDP and employs over 70,000 highly skilled jobs.[146]

However, the continued, inclusive growth of Auckland's tech industry faces three main challenges[147]:

- Spatial concentration, where most of the larger, high-performing tech firms are situated in three Local Board areas, i.e., Waitematā, Maungakiekie-Tāmaki, and Upper Harbour.
- Skills shortages and mismatches, due to an over-dependence on sourcing labour from overseas and the shrinking offshore talent tool intensified by COVID-related travel restrictions.
- Lack of diversity in the workforce, which is predominantly represented by males, 30–49 age group, and European or Asian ethnic groups.

To overcome these challenges, the Tech Tāmaki Makaurau strategic programme sets out the vision to become "a global epicentre of innovation, technology and talent that creates an inclusive and equitable future for

[143] Auckland Unlimited. (n.d.). *Industry*. https://aucklandunlimited.com/portfolio/industry.
[144] Auckland Unlimited. (25 February 2022). *Tech Tāmaki Makaurau*. https://industry.aucklandnz.com/business/techakl.
[145] *Ibid.*
[146] *Ibid.*
[147] *Ibid.*

all".[148] Under this vision, three goals were established for Auckland's tech industry to drive economic prosperity for the communities[149]:

- Creating a tech ecosystem that is a beacon for home-grown and international talent.
- Developing a tech industry that delivers a sustainable and equitable future for all.
- Growing Tāmaki Makaurau as a globally connected, inclusive and collaborative tech city.

Accordingly, a set of actions were identified[150]:

- Increase employment in the tech industry, especially for unrepresented population groups, e.g., Māori, Pacific peoples, youth, and women.
- Expand and scale tech career pathway programmes.
- Attract tech talent at various levels from overseas.
- Collaborate with other stakeholders to advance equity in accessing technology and resources.
- Support social procurement made by all local government organisations.
- Reduce inequity of funding access for indigenous tech businesses.
- Create knowledge-intensive hubs across Auckland.
- Identify and promote thriving tech leaders and companies for domestic and international visibility.
- Expand the industry size of export-ready tech businesses.
- Establish partnership agreements with overseas tech hubs.

To ensure successful delivery of Tech Tāmaki Makaurau, Tātaki Auckland Unlimited develops and implements collaborative partnership arrangements across government agencies and with industry. A new tech

[148] *Ibid.*
[149] *Ibid.*
[150] *Ibid.*

industry advisory group is established to guide the implementation of the programme and represent the collective voice of Auckland's tech industry.[151]

2.4 Implementation and Impact of Smart Growth Strategy

Over the past decade, Auckland has implemented smart growth strategy in both the private and public sectors to create an innovative tech ecosystem. This section elaborates on the implementation and impact of Auckland's smart growth strategy through two case studies: GridAKL innovation hubs focusing on the private sector and digital transformation in the government sector.

2.4.1 *GridAKL Innovation Hubs*

Located in the heart of Auckland's Wynyard Quarter Innovation Precinct, the GridAKL innovation hubs are delivered and managed by Tātaki Auckland Unlimited (the economic development agency of Tātaki Makaurau Auckland) on behalf of the Auckland Council.[152] The GridAKL innovation hubs aim to cultivate innovation by creating a collaborative community of purpose-driven technology and innovative businesses.[153] The role of GridAKL is to support high-impact, growth-orientated, technology-focused businesses and entrepreneurs, enabling them to develop and commercialise their innovative solutions.[154] GridAKL adopts a co-working space model to develop its physical space, which creates pathways for innovative start-ups and small- and medium-sized enterprises (SMEs) to connect and share ideas. It also provides various resources,

---

[151] *Ibid.*

[152] Auckland Unlimited. (n.d.). *Helping Auckland Business*. https://aucklandunlimited.com/helping-auckland-businesses.

[153] Bridley M. (4 July 2022). Interview with the Head of Technology and Skills, GridAKL [Personal communication].

[154] GridAKL. (n.d.a). *About Us*. https://gridakl.com/about-us/.

such as legal advice, investment, and business support, that innovative start-ups and SMEs need to grow sustainably.[155]

The GridAKL programme was started in 2015 with a NZ$30 million investment from Auckland Council — the largest investment in the economic development of its time.[156] The initial project is GridAKL/John Lysaght (Figure 2.1). As of April 2024, this co-working space houses 42 purpose-driven start-ups and SMEs in the innovative industry.[157] Key sectors include software services (e.g., BoardPro and Canva), AI-powered data analytics (e.g., Thematic), game development (e.g., Geo AR and Kids Game Dev), impact evaluation and consultancy for B Corps (e.g., B Lab), and niche companies providing specialised services and advice for growing innovative industries (e.g., The Connective and Albatross PR).[158]

In 2017, the GridAKL innovation programme was scaled up, building a second innovation hub, 12 Madden Street (Figure 2.2). Sharing a common goal with John Lysaght, the 12 Madden Street space seeks to create an innovation ecosystem and attract talent to work, live, and invest in Auckland.[159] This hub provides shared workstations, technology facilities, and innovation platforms to 83 businesses,[160] some of which are spin-offs from the GridAKL/John Lysaght.[161] The 12 Madden Street innovation hub has also attracted multinational companies, such as IBM and HP, to set up their regional offices. Key sectors based in the 12 Madden Street innovation hub include IT services (e.g., Snowflake and Equinox IT), digital service design (e.g., Foundry and ThinkPlace), information, media, and technology (e.g., Spark and Mi9), professional services (e.g., BBG and TEDxAuckland), and tech for social impact (e.g., AucklandPride and Bluetree).[162]

---

[155] *Ibid.*

[156] Bridley (4 July 2022). *Op. cit.*

[157] GridAKL. (n.d.b). *John Lysaght.* https://gridakl.com/our-spaces/lysaght-building/.

[158] GridAKL. (n.d.c). *GridAKL/John Lysaght Residents.* https://gridakl.com/our-residents/?_sft_location=john-lysaght-building.

[159] Bridley (4 July 2022). *Op. cit.*

[160] GridAKL. (n.d.d). *GridAKL/12 Madden Street Residents.* https://gridakl.com/our-residents/?_sft_location=12-madden-st.

[161] Bridley, M. (4 July 2022). *Op. cit.*

[162] GridAKL. (n.d.d) *Op. cit.*

Figure 2.1. GridAKL/John Lysaght.

Source: Authors.

Figure 2.2. GridAKL/12 Madden Street.

Source: Authors.

In 2022, a third space in the GridAKL innovation ecosystem, GridMNK, was launched in Manukau City in South Auckland to scale up the Tātaki Makaurau innovation ecosystem, contributing to a more connected, equitable, and thriving economy (Figure 2.3).[163] GridMNK aims to provide economic and upskilling opportunities to make generational change in south and west Auckland where the population is of lower prosperity and to support Maori and Pasifika entrepreneurs.[164]

GridAKL innovation hubs won the 2019 Economic Development New Zealand (EDNZ) Awards for Innovation, demonstrating Auckland's leadership role in developing smarter solutions for complex urban, economic, and sustainability challenges.[165] The governance model of GridAKL

---

[163] *Ibid.*

[164] *Ibid.*; GridMNK. (n.d.e). *About Us*. https://www.gridmnk.nz/about.

[165] Our Auckland. (2019). *Auckland Wins Recognition for Public Innovation*. https://ourauckland.aucklandcouncil.govt.nz/news/2019/11/auckland-wins-national-recognition-for-public-innovation/.

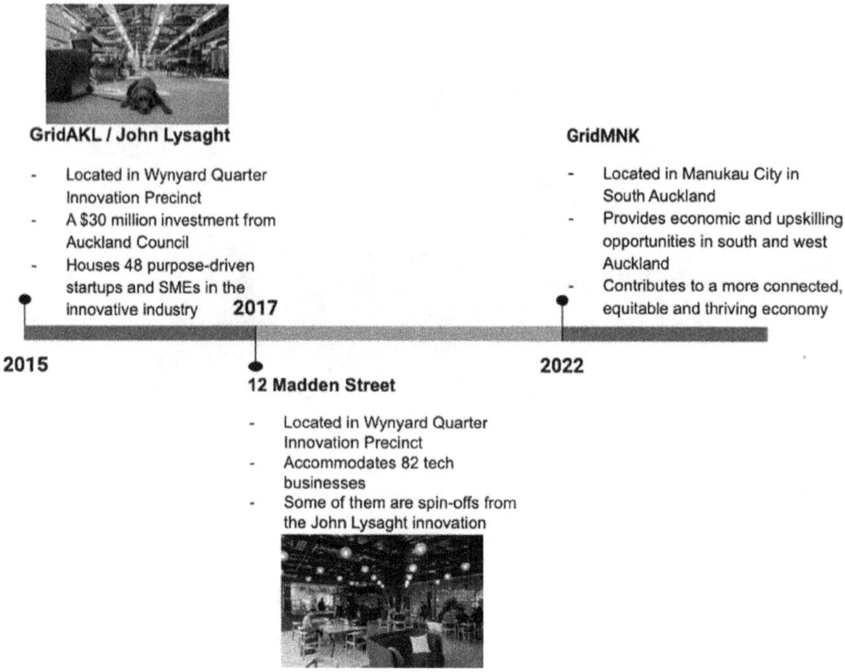

Figure 2.3. Development timeline of innovation hubs in Auckland.
Source: Adapted from GridAKL.[166]

also exemplifies how an economic development agency, Tātaki Auckland Unlimited in this case, successfully functioned as a catalyst for long-term systemic changes in the innovation ecosystem, generating significant benefits for both the general public and the private companies engaged in the GridAKL programmes.[167]

Some tech and innovative start-ups at GridAKL have been recognised for their innovative actions towards smart growth. For instance, clean tech start-up, For The Better Good, has received the 2020 Impact Award for Climate for its leadership and innovative action to support Auckland's transition to a more resilient, low-carbon economy and society. This clean tech company produced New Zealand's first biodegradable and compostable water bottle made from plants. It also completely re-designed the

[166] GridAKL. (n.d.a). *Op. cit.*
[167] *Ibid.*

production system to reduce carbon emissions and waste and sequester carbon.[168]

Critical success factors for GridAKL innovation hubs include the following[169]:

- **Location**: Two of the GridAKL innovation hubs (John Lysaght and 12 Madden Street) are strategically located in the centre of Wynyard Quarter Innovation Precinct, facilitating agglomeration of tech and innovation industries.
- **Public–private partnership**: The development of GridAKL was initiated by a NZ$30 million investment from Auckland City Council. Its operator, Auckland Unlimited, has taken long leases (12 years) on buildings at GridAKL to encourage private developers to bring the buildings forward. The long leases give the private developers confidence and reduce their risk in investment.
- **Community building**: Creating a community of purpose-driven tech and innovation start-ups and SMEs is key for founders and entrepreneurs to connect and scale up. This community also involves a range of "anchor tenants" that are large enough to support smaller businesses.
- **Continuous support to sustainable growth**: GridAKL innovation hubs provide key business support to assist tech and innovative start-ups and SMEs to grow sustainably. This approach differs from a typical accelerator or incubator approach to growing start-ups and SMEs through a 3- to 12-week programme, after which the support is turned off.
- **An urbanised, playful approach to physical development**: GridAKL provides co-working space in the innovation hubs to facilitate innovation spill-overs and collaboration among like-minded tech and innovative start-ups and SMEs. Key amenities and residential apartments are also developed to attract people to work, live and play in the area. This approach deliberately shifts away from the traditional suburban business park approach.

---

[168] GridAKL. (2020). *Our Stories. Jayden Klinac Winner of the Impact Award for Climate 2020.* https://gridakl.com/jayden-klinac-winner-of-the-impact-award-for-climate-2020/.
[169] Bridley (4 July 2022). *Op. cit.*

In advancing GridAKL innovation hubs, there are several key challenges, including the following[170]:

- **Addressing community needs and demands**: This is critical to building a community of innovation businesses. Research and learning from the community are essential to understanding their needs and guiding the innovation hub in its provision of spaces, services, and tools needed to grow the innovation ecosystem.
- **Purpose-driven**: The growth of the tech and innovation industry is not static but needs to evolve with time and purpose to remain relevant, e.g., become more diverse, create pathways into the industry for other communities, and essentially focus on tech-for-good and tech-solving problems.
- **Having the right operator**: Every innovation hub is unique in its location and its community of tech and innovation businesses. The right operator is a critical factor in delivering the daily operation, understanding the industry and the needs of the community, and identifying the focus of the innovation hub, e.g., clustering towards a particular niche sector.

2.4.2 *Digital Transformation in Auckland's Public Sector*

Another key area of digital economy development in Auckland is the digital transformation of the city's public sector. As revealed by the New Zealand Technology Industry Association, the public sector is the largest customer of the technology industry in New Zealand, with the combined spending across central and local governments on ICT accounting for 40% of the tech sector's domestic revenue.[171] The uptake of digitalisation across government agencies not only enhances public service delivery but also stimulates innovation and productivity for the tech sector. As recognised in the 2015 Business Growth Agenda, New Zealand's transition to a

---

[170] *Ibid.*

[171] New Zealand Technology Industry Association (NZTech). (2016). *From Tech Sector to Digital Nation*. https://nztech.org.nz/wp-content/uploads/sites/8/2019/02/from-tech-sector-to-digital-nation-2nd-edition-ebook.compressed.pdf.

leading digital nation requires a shared effort from a variety of stakeholders in the digital community, including the technology sector, traditional business sectors, industry associations, the public, and government.[172]

Auckland Council has been pursuing digital transformation of government services, products, and processes in recent years. It established the digital services strategy in 2019, aiming to connect citizens to government services in a personalised and low-friction way.[173] This strategy takes a people-centric approach to digitalising public services and products, rather than just improving government ICT system efficiency.[174] Key policies and plans are accessible to all via digital platforms, such as the Auckland Plan 2050[175] and the Auckland City Centre Master Plan 2020.[176] These platforms illustrate the vision and specific areas of the plan through a suite of well-coordinated websites, instead of listing the policy document as a single web link for download. The digital plans are easy to navigate with the use of indicative visuals and prompts (Figure 2.4). This provides citizens with a good overview of what the city could be like in the future and also allows them to locate and access the information needed for specific purposes.

In addition to digital services, Auckland Council has developed digital products and data analytic tools to improve citizens' decision-making and quality of life. For instance, Auckland Council developed the Safeswim website in partnership with other regional councils, including Surf Life Saving New Zealand, Surf Life Saving Northern Region, and the Auckland Regional Public Health Service. The Safeswim digital platform provides real-time advice on water quality and safety risks for swimming at specific locations across the Auckland region, with the aim to enable beach users to make more informed decisions about when and where to swim safely (Figure 2.5).[177] Safeswim integrates real-time data on wastewater

---

[172] MBIE (2015). *Op. cit.*

[173] Auckland Technology and Governance Team. (1 July 2023). Interview with Auckland technology and governance team [Personal communication].

[174] *Ibid.*

[175] Auckland Council. (2018). *Op. cit.*

[176] Auckland Council. (2020). *Auckland City Centre Master Plan 2020.* https://www.aucklandccmp.co.nz.

[177] Swimsafe. (n.d.). *Decide with Swimsafe.* https://safeswim.org.nz.

Figure 2.4. Digital platform of Auckland City Centre Master Plan 2022.
Source: Auckland Council.[178]

and stormwater networks (tides, wind speeds and direction, rip currents, jellyfish, and other physical hazards) into predictive models to forecast water quality and levels of risk associated with swimming across various beach sites.[179] The use of water sensors and predictive models significantly improves the accuracy of water quality, safety hazards, and public health risks, compared with the traditional approach of collecting water samples.[180] Digital signages are also installed at selected beaches to inform beach users without access to smartphones or the internet.[181]

Auckland Council's achievement in digital transformation is recognised internationally. The Safeswim programme won the 2018 Smart Cities Asia-Pacific Award for Smart Water Category following an

[178] *Ibid.*

[179] *Ibid.*

[180] Swimsafe. (2022). *Safeswim's Water Quality Programme*. https://education.safeswim.org.nz/articles/safeswim-water-quality-programme.

[181] OpenGov. (2019). *Auckland, New Zealand Recognised for Smart City Innovations*. https://opengovasia.com/auckland-new-zealand-recognised-for-smart-city-innovations/.

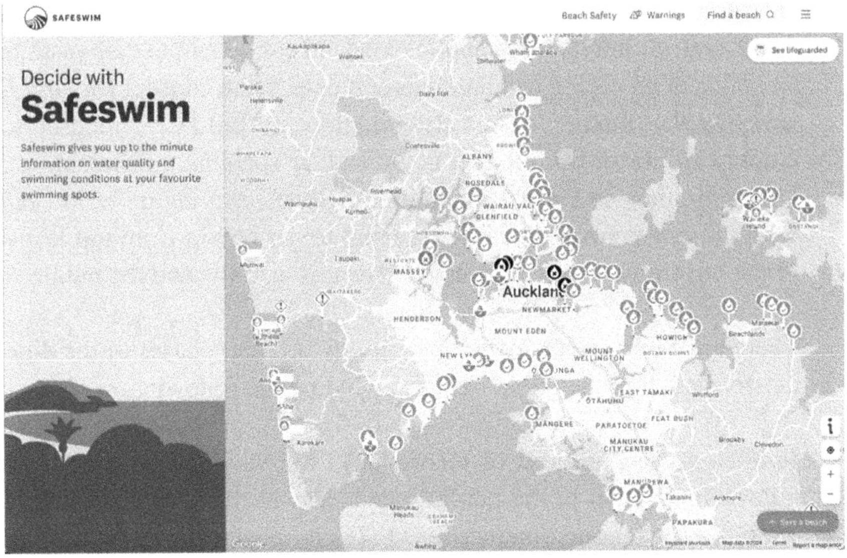

Figure 2.5. Safeswim platform.
Source: Safeswim.[182]

intensive six-month process of monitoring and evaluating hundreds of smart city initiatives in the Asia-Pacific region. It demonstrates how the local government collaborates effectively with multiple stakeholders to deliver tech-for-good.[183] Reflecting on its digital transformation journey, Auckland Council identified several critical success factors[184]:

- **A clear understanding of the capacity of technology**: Avoid overselling what the technology can do. The blue-sky thinking of digital technology often happens at the highest levels of the organisation, without a thorough consideration of the practical implementation. A realistic implementation plan is key to minimising the expectation-deliverable gap.
- **Deliberate learning from failed projects and misunderstanding**: The digital transformation journey inevitably involves the use of

[182] Swimsafe. (n.d.). *Op. cit.*
[183] Our Auckland. (2018). *Safeswim Receives 2018 Smart Cities Award.* https://ourauckland.aucklandcouncil.govt.nz/news/2018/7/auckland-councils-safeswim-receives-asia-pacific-gong-in-idcs-2018-smart-cities-awards/.
[184] Auckland Technology and Governance Team. (1 July 2023). *Op. cit.*

emerging technologies, which are subject to uncertainties. Through small-scale prototypes and applications, it is important to demonstrate the value of emerging technologies and de-risk the concept for larger-scale implementation. This also requires a "fail-fast" mindset to quickly fail, learn from it, and go on and try something different.
- **Obtaining buy-in from stakeholders**: Upon verifying the value and feasibility of innovations, it is essential to get buy-in from top levels of the government, other agencies, business partners and the public.

There are challenges to overcome during different phases of the digital transformation in the public sector, including the following:

- **Realising the potential of technology**: Before implementing technology innovation in the public sector, it is challenging to convince various agencies and organisations to realise how some of the technological developments can deliver value for them. Regular and clear communication channels with all stakeholders is central to technology adoption.
- **Tackling uncertainties**: When implementing a new technology around a new field, it is challenging to manage the uncertainties with the innovation. Experimentation and the freedom to experiment are key strategies.
- **Funding**: It is critical to access and secure funding to keep the momentum up while testing new technology and innovative solutions. Under the disruptive impacts of the COVID-19 pandemic, it has become less of a priority for the local government to fund some of the innovation projects on emerging technologies.

2.5 Conclusion

As the economic powerhouse and innovation hub of New Zealand, Auckland plays a leadership role in driving smart growth both regionally and nationally. Auckland has been pursuing innovative solutions for digital transformation in both the private and public sectors, representing a multifaceted approach to smart growth. A well-aligned policy framework is essential. In line with the national strategies on digital economy, the

Auckland Council establishes strategic directions for smarter and more inclusive economic development in the city's strategic plan (Auckland Plan 2050). These strategies are further operationalised in a domain-specific programme, Tech Tāmaki Makaurau (2022–2025), to drive the growth of the technology industry and attract tech talent and investment in Auckland. This strategy focuses on addressing challenges facing Auckland's tech sector, such as skill shortages, spatial inequality, and lack of diversity in the tech workforce. A set of specific actions are identified to build the ecosystem of tech and innovation industries in Auckland.

A governance model of innovation hubs is applied and scaled up. Auckland takes a place-based approach to developing innovation hubs within the city's broader innovation precinct. The economic agency Tātaki Auckland Unlimited not only operates the GridAKL innovation hubs but also provides key services and advice to tech and innovation start-ups and SMEs to help them grow sustainably. This approach differs from the typical accelerator/incubator model, which often has a shorter runway. The initial funding from the Auckland Council and the continued support from Tātaki Auckland Unlimited functioned as a catalyst to bring out systemic changes in the innovation ecosystem for the long term.

Community building is another integral part of smart growth. Auckland's innovation hubs seek to create a community of purpose-driven tech and innovation start-ups and SMEs. The co-working spaces enable funders and entrepreneurs to connect with each other and scale up. The onsite support from Tātaki Auckland Unlimited is also key to creating an innovation community. With the city's strategic plan for more inclusive economic growth, an innovation hub was launched in South Auckland to accelerate upskilling and job creation for innovative industries in the area.

Public sector is an active player in the innovation system. As the single largest consumer of the technology industry, Auckland Council has taken an active role in developing and implementing digital solutions for enhanced delivery of public services. Strategic plans and other key policies are accessible to all via digital platforms on the council website. Digital products, like the Safeswim platform, are developed to help citizens make informed decisions and improve public health associated with water quality. Digital signages are in place at selected sites to overcome the issue of not having access to a smart device or the internet.

CHAPTER 3

GOTHENBURG: SMART TOURISM

3.1 City Overview

Location	Gothenburg, Sweden
Land Size	447 km^2
Climate	Marine west coast climate
Geography (topology/landscape)	Coastal city
Environmental Risks	Climate change
Population	596,841 (2022)
GDP	US$43.7 million (2021)
GDP Per Capita	US$74,568 (2021)
Key Urban Issues of Smart City	Carbon footprint of over-tourism, socio-spatial segregation, income inequality

Gothenburg is Sweden's second biggest city and a city municipality, collectively known as the City of Gothenburg. The city is located on the western coast of Sweden, midway between Copenhagen, Denmark, and Oslo, Norway, spanning over 447 km^2.[185] The city is encircled by a deep forest, tranquil lakes, rocky shoreline, and the sea. Gothenburg has a

[185] World Population Review. (n.d.). *Gothenburg Population 2023*. https://worldpopulationreview.com/world-cities/gothenburg-population.

marine west coast climate.[186] Temperatures are mild in summer and cold in winter.[187] The mean temperature in Gothenburg is 8.1°C, where the lowest temperature is in February at –0.3°C while the highest is in July at 17.3°C.[188] The city experiences a total of 998 mm of rainfall per year, with the greatest amount of precipitation occurring in August and the lowest in March.[189] The city is preparing to face a sea level rise of 1 m by 2100 in addition to an increased frequency of flooding events.[190]

In 2022, Gothenburg had a population of 596,841 and a population density of approximately 1,300 per square kilometre.[191] The city has a young population (approximately 55% of the population is below 40 years old). As of 2021, approximately three-quarters of the people living in Gothenburg were born in Sweden. Over 140,000 immigrants live in Gothenburg, with the majority from Iraq, Iran, Somalia, Bosnia, Herzegovina, Finland, and Poland, among others. Population ageing is a major demographic development in Sweden. Despite this trend, population ageing has been less pronounced in Gothenburg than in many other regions of Sweden.[192] In 2018, the percentage of women

---

[186] Carlsen, H. K., Oudin, A., Steingrimsson, S., & Oudin Åström, D. (2019). Ambient temperature and associations with daily visits to a psychiatric emergency unit in Sweden. *International Journal of Environmental Research and Public Health, 16*(2), Article 2. https://doi.org/10.3390/ijerph16020286.

[187] Climates to travel. (n.d.). *Climate — Gothenburg (Sweden).* https://www.climatestotravel.com/climate/sweden/gothenburg.

[188] Climate Data. (n.d.). *Gothenburg Climate.* https://en.climate-data.org/europe/sweden/vaestra-goetalands-laen/gothenburg-197/.

[189] *Ibid.*

[190] International Water Association. (n.d.). *Gothenburg.* https://iwa-network.org/city/gothenburg/.

[191] Official Statistics of Sweden. (n.d.a). *Population Statistics Number of Persons by Region, Population Changes, Period, Sex and Year.* https://www.statistikdatabasen.scb.se/pxweb/en/ssd/START__BE__BE0101__BE0101G/BefforandrKvRLK/table/tableViewLayout1/; Business Region Goteborg. (2022). *Gothenburg Demographics.* http://locate.investingothenburg.com/.

[192] Mats Stjernberg. (2019). *ACPA — Adapting European Cities to Population Ageing: Policy Challenges and Best Practices.* https://www.espon.eu/sites/default/files/attachments/8.%20ACPA_city%20report_Gothenburg_1.pdf.

(55.3%) was slightly higher than the percentage of males (44.7%) in the population aged 65 and over in Gothenburg. The gender demographic reflects the general situation throughout Sweden. In 2019, the average life expectancy for men is 79 years and 83 years for women. A notable distinction between older men and women in Gothenburg is the household structure. In 2017, around 50% of older women in Gothenburg lived alone, compared to approximately 30% of older men. This conforms to the general situation in Sweden. In Gothenburg, the districts where older people constitute more than 25% of the total population are primarily found in the city's southern and southwestern suburbs, whilst the districts in the inner city have significantly younger age profiles.

Gothenburg has successfully transitioned from an industrial-based economy to a knowledge-intensive one, demonstrating sustainable growth.[193] The city's regional economy is characterised by international industrial enterprises, such as Volvo, SKF, and AstraZeneca. Gothenburg's economy has gradually expanded industrial investments and exports and imports growth. Since 2010, the business environment in the Gothenburg region has experienced a rapid increase in exports, with a boom period from 2014 to 2018.[194] Research and development are another of Gothenburg's economic pillars, with a cutting-edge manufacturing industry and strong collaboration between businesses, politicians, and academics, ensuring a high level of competence and innovation capabilities.[195] In 2021, Gothenburg's gross regional domestic product (GRDP)

---

[193] Xu, X. & Ström, P. (2016). The transformative roles of knowledge-intensive business services in developing green ICT: Evidence from Gothenburg, Sweden. In A. Jones, P. Ström, B. Hermelin, & G. Rusten (Eds.), *Services and the Green Economy* (pp. 99–124). UK: Palgrave Macmillan.

[194] Business Region Göteborg AB. (n.d.). *100% Facts & Figures* (p. 21). https://www.investingothenburg.com/sites/investingothenburg/files/downloadable_files/brg_fact_and_figures_2022_web.pdf.

[195] *Ibid.*; Swedish Exhibition & Congress Centre Group. (n.d.). *The Gothenburg Region. The Swedish Exhibition & Congress Center*. https://en.svenskamassan.se/explore-gothenburg/business/.

was 435.99 billion SEK[196] with a GRDP per capita of 745,000 SEK.[197] The region is presently witnessing an economic boom following a rapid recovery from the COVID-19 pandemic.[198]

Gothenburg's labour market remains strong.[199] Sectors, such as hospitality, information and communication, transport, knowledge-intensive services, and manufacturing see the biggest increases in employment.[200] Gothenburg's strategic location, expanding infrastructure, new logistic operations in freight logistics, warehousing, and terminal management, and multiple investments in the port provide ideal conditions for future growth.[201] In 2023, the Gothenburg region was named Sweden's best logistics location by the Swedish trade journal, *Intelligent Logistik*, and the top three large European cities of the future for human capital and lifestyle by the 2023 fDi Intelligence's European Cities and Regions of the Future.[202] Gothenburg also ranked 8 and 10 for the categories fDi index and economic potential respectively, highlighting that the city is an attractive location for international firms to grow and innovate.

---

[196] Official Statistics of Sweden. (n.d.b). *Gross Regional Domestic Product (GRDP, ESA2010), Current Prices, Million SEK by Region and Year. PxWeb*. https://www.statistikdatabasen.scb.se/pxweb/en/ssd/START__NR__NR0105__NR0105A/NR0105ENS2010T06A/table/tableViewLayout1/.

[197] Official Statistics of Sweden. (n.d.c). *Gross Regional Domestic Product (GRDP, ESA2010) Per Capita, Current Prices, Thousand SEK by Region and Year*. https://www.statistikdatabasen.scb.se/pxweb/en/ssd/START__NR__NR0105__NR0105A/NR0105ENS2010T06A/table/tableViewLayout1/.

[198] Business Region Goteborg. (2021). *Economic Outlook* (p. 7). https://www.investingothenburg.com/sites/investingothenburg/files/downloadable_files/economic-outlook-3-2021.pdf.

[199] Invest in Gothenburg. (2023a). *Economy Slows But Labour Market Remains Strong*. https://www.investingothenburg.com/news/all-news/economy-slows-labour-market-remains-strong.

[200] *Ibid*.

[201] Invest in Gothenburg. (2023b). *Infrastructure and Logistics*. https://www.investingothenburg.com/taxonomy/term/965.

[202] Intelligence, fDi. (16 February 2023). *European Cities and Regions of the Future 2023 — The Winners*. https://www.fdiintelligence.com/content/rankings-and-awards/european-cities-and-regions-of-the-future-2023-the-winners-82056.

The local government is made up of an elected City Council, comprising 81 members and 44 deputies, acting as the highest decision-making body in Gothenburg.[203] Elections occur once every four years. For the City Council to make decisions, more than half of its members must be present.[204] The City Council also elects municipal councillors responsible for their respective areas of activity. The city has several municipal companies, e.g., Goteborg & Co is a municipality-owned company with a mission to market and support the development of Gothenburg as a city for tourism, events and meetings.[205] Göteborgs Stadshus AB is the parent company for the companies owned by the city.[206]

3.2 Smart City Development: Multifaceted Sustainability

Gothenburg began developing sustainability-focused strategies in 2001.[207] Within the Gothenburg Metropolitan Area, the idea of sustainability acted as a cohesive and empowering narrative, inspiring a coalition of technology and economics experts to engage proactively and govern the city region effectively. This enabled the initiation and implementation of critical infrastructure necessary for driving forward economic growth projects.[208] The city's sustainability-focused strategies integrate various technologies and initiatives to create a resilient and sustainable urban environment that prioritises efficient resource use and an improved quality of life for its residents.

---

[203] Göteborg Stad. (n.d.a). *Om kommunfullmäktige.* https://goteborg.se/wps/portal?uri=gbglnk%3agbg.page.e93ce75d-fbca-4998-bc87-ad5599c8d6c8.

[204] *Ibid.*

[205] Göteborg & Co. (n.d.a). *Organisation.* https://goteborgco.se/en/about-goteborgco/organisation/.

[206] Göteborg Stad. (n.d.b). *Göteborgs Stadshus AB.* https://goteborg.se/wps/portal?uri=gbglnk%3a20140519094420.

[207] Göthenburg & Co. (2016). *Gothenburg 2021* (p. 228). https://www.goteborg2023.com/uploads/2016/10/Proposed-work-plan-goteborg2021_120530-1.pdf.

[208] Tahvilzadeh, N., Montin, S., & Cullberg, M. (2017). Functions of sustainability: Exploring what urban sustainability policy discourse "does" in the Gothenburg Metropolitan Area. *Local Environment*, *22*(sup1), 66–85. https://doi.org/10.1080/13549839.2017.1320538.

3.2.1 *Sustainability Development Visions*

The city's approach to sustainable development is guided by its strategic visions and long-term goals. Some of the projects and initiatives to promote social and environmental sustainability, and eco-friendly solutions for a resilient urban future, include the following:

- Gothenburg 2050 was initiated in 2001 with the goal of accelerating development towards a sustainable future city. Through the project, long-term visions were developed to aid the city's goals. There are five primary visions in the Gothenburg 2050 project[209]:
  - Solar city Gothenburg 2050 — strategic energy planning for a sustainable energy system.
  - Transportation Gothenburg 2050 — long-term sustainable transport system.
  - Urban design Gothenburg 2050 — sustainable urban design for suburban life and at the local square.
  - Eco-cycling Gothenburg 2050 — sustainable water and waste management.
  - Food Gothenburg 2050 — long-term strategies for sustainable food and grocery chains.
- River City Gothenburg — Open to the World (2012) follows three primary strategies centred around improving social sustainability goals and reinforcing the regional core to improve the local environment for the entrepreneurs and businesses in the city.[210] While the River City vision and related strategies were gradually implemented, the project was criticised for its lack of performance measurements, articulated targets, and organisation output at the governance level.[211]
- Gothenburg 2016 City Strategy "Equal Gothenburg" combines the aim of equality and social sustainability to develop an equal city.

---

[209] Phdungsilp, A. (2011). Futures studies' backcasting method used for strategic sustainable city planning. *Futures*, *43*(7), 707–714.

[210] Brorström, S., Argento, D., Grossi, G., Thomasson, A., & Almqvist, R. (2018). Translating sustainable and smart city strategies into performance measurement systems. *Public Money & Management*, *38*(3), 193–202.

[211] *Ibid.*

The strategy aims to reduce disparities in living conditions, making cities inclusive, safe, healthy, and socially resilient. The city's support for the sharing economy is primarily motivated by the promotion of social integration, equity, justice, and social capital.[212] The strategy consists of four key areas:

- Establish a successful start in life and good conditions during schooling years.
- Establish requirements for employment.
- Establish health-promoting and sustainable environments and communities.
- Create preconditions for participation, influence, and trust.

3.2.2 *Innovative Urban Planning*

Gothenburg is working on innovative urban planning to produce more sustainable, efficient, and liveable urban environments. This is done by integrating physical and digital aspects, utilising real-time data, and emphasising energy efficiency and environmental sustainability in urban living spaces. Innovative urban planning projects and initiatives with public life at the heart of their vision include the following:

- Strategies towards Energy Performance and Urban Planning (STEP UP) (2012–2015) is an EU-funded project. Via STEP UP, Gothenburg developed two innovative pipeline projects — smart urban logistics and planning for sustainable urban lifestyles.[213] STEP UP also guided Gothenburg to enhance its Sustainable Energy Action Plan (SEAP).[214] Gothenburg's energy plan (Climate Programme) was successfully launched in 2014. The Climate Programme builds on several earlier climate actions and includes objectives and strategies to achieve an

[212] Goteborg Stad. (n.d.c). *Equal City — Making the Entire City Socially Sustainable.* https://goteborg.se/wps/portal?uri=gbglnk%3a201781145134969.

[213] STEP UP. (n.d.). *Pipeline Projects.* https://www.stepupsmartcities.eu/CityProjects/PipelineProjects/tabid/5065/Default.aspx.

[214] Smart Cities Marketplace. (n.d.a). *STEP-UP.* https://smart-cities-marketplace.ec.europa.eu/projects-and-sites/projects/step.

equitable and sustainable level of greenhouse gas emissions by 2050.[215] The programme involves developing large-scale production of renewable electricity and reducing consumption of resource-intensive goods.

- Combined Efficient Large Scale Integrated Urban Systems (CELSIUS) (2013–2017) is an EU-funded project that aims to connect a smart heating system to 900 apartments in high-density city areas of Gothenburg, connect docked ships to district heating, and transfer cold water from the river into the district cooling network.[216]
- The European Cities serving as Green Urban Gate towards Leadership in sustainable Energy (EU-GUGLE) (2013–2018) is an EU-funded project. The project consists of eight cities including Gothenburg to share research efforts relevant to sustainable renovation in urban areas to catalyst large-scale Europe-wide replication in smart cities and communities by 2020.[217] The project seeks to help Gothenburg address the following issues: improve the housing conditions in areas of Hammarkullen where families often live in overcrowded conditions, discover ways and finance models to renovate the housing stock without significant rent increases, and improve aesthetic features of buildings and outdoor spaces.[218]
- Integrated and Replicable Solutions for Co-Creation in Sustainable Cities (IRIS) (2017–2022) is a Lighthouse project, which consists of three European cities: Utrecht, Nice and Gothenburg. The EU-funded project addressed areas in smart energy management, smart e-mobility, an innovation platform for digital cities, residents' involvement, and co-creation.[219] Gothenburg's IRIS solutions focus on

---

[215] City of Gothenburg. (2015). *Climate Programme for Gothenburg*. https://carbonn.org/uploads/tx_carbonndata/Climate%20Programme_%20Folder.pdf.

[216] Smart Cities Marketplace. (n.d.b). *CELSIUS Site Gothenburg*. https://smart-cities-marketplace.ec.europa.eu/projects-and-sites/projects/celsius/celsius-site-gothenburg.

[217] Smart Cities Marketplace. (n.d.c). *EU-GUGLE*. https://smart-cities-marketplace.ec.europa.eu/projects-and-sites/projects/eu-gugle.

[218] Sinfonia. (n.d.). *Gothenburg, SE*. http://www.sinfonia-smartcities.eu/en/replication-cities/gothenburg.

[219] Smart Cities Marketplace. (n.d.d). *IRIS Gothenburg*. https://smart-cities-marketplace.ec.europa.eu/projects-and-sites/projects/iris/iris-gothenburg; IRIS Smart Cities. (16 March 2023). *Co-creating Smart and Sustainable Cities*. https://irissmartcities.eu/.

testing innovative energy management and storage to establish energy-positive districts. Collaborative efforts of authorities, service providers and citizens of Gothenburg helped the city to tackle urban sustainability concerns through several approaches such as open data, open innovation, and community engagement.[220]

- **Smart City Sweden**: In 2020, the Swedish Energy Agency announced the expansion of Smart City Sweden. Smart City Sweden is a state-funded initiative and national demonstration platform for smart and sustainable city solutions. One of the six regional Smart City Sweden offices is established in Gothenburg to boost the opportunity for international delegations to experience smart and sustainable solutions located in Sweden.[221] The scope of solutions under the Smart City Sweden platform includes Energy, Climate and Environment, Mobility, Social sustainability, and Urban planning and Digitalisation.
- **Virtual Gothenburg**: As part of Gothenburg's smart transition, a digital twin, a 3D model of Virtual Gothenburg, was launched in 2021. Virtual Gothenburg contains a large amount of urban data and aims to lay the foundation for a more efficient and smart urban planning process and as a testbed for innovations.[222] The initiative supports detailed studies of the city from three perspectives[223]:
  - understanding what the existing city looks like and how it works,
  - controlling different functions in the city based on what is happening in real-time,
  - predicting and planning simulated future events in the city.
- **100 Climate-Neutral and Smart Cities**: In April 2022, Gothenburg became one of the European Union's 100 Climate-Neutral and Smart

[220] Johanneberg Science Park. (n.d.). *Activities, Articles and Videos*. https://www.johannebergsciencepark.com/en/projects/iris-smart-cities/activities-articles-and-videos.

[221] IVL Swedish Environmental Research Institute. (2019). *Smart City Sweden is Expanding*. https://www.ivl.se/english/ivl/press/press-releases/2019-01-09-smart-city-sweden-is-expanding.html.

[222] Visual Arena. (n.d.). *What Do You Want to Test?* https://visualarena.lindholmen.se/en/virtual-gothenburg-lab/vad-vill-du-testa.

[223] Göteborgs Stad. (2021). *Digital Twin*. https://stadsutveckling.goteborg.se/sa-planeras-staden/digitaltvilling/digital-twin/.

Cities.[224] As a member of the network, Gothenburg would receive guidance and support to accelerate its climate-related efforts to achieve climate neutrality by 2030.[225]

3.2.3 *Sharing Economy*

Gothenburg has been nurturing the sharing economy, an economic model that promotes resource efficiency and community collaboration. The city has established testbeds for a variety of sharing services and digital solutions including the following:

- Gothenburg's 2015 budget supported the growth of the sharing economy. In the budget report, the city encouraged the development of sharing economy "to reduce environmental impacts, increase process efficiency and reduce costs." In 2017, the Consumer and Citizen Service Administration at the Gothenburg municipality presented a vision for Gothenburg to become a Sharing City and received support from citizens.[226]
- Sharing Cities Sweden (2017–2021) is a national programme as part of the Viable Cities, a Swedish innovation programme for climate-neutral cities by 2030, which is funded by Swedish Innovation Agency VINNOVA, Swedish Energy Agency and FORMAS.[227] Through Sharing Cities Sweden, Gothenburg, Stockholm, Malmö, and Umeå actively engaged with the sharing economy to develop world-leading testbeds. The testbeds created the conditions for sharing services and digital solutions to be developed, tested, and evaluated.[228] The city coordinated the Sharing City Gothenburg testbed and engaged a total of 15 local partners from the public sector, business, academia, and

---

[224] Eurocities. (29 April 2022). *The 100 Climate-neutral and Smart Cities by 2030*. https://eurocities.eu/latest/the-100-climate-neutral-and-smart-cities-by-2030/.

[225] Net Zero Cities. (n.d.). *Towards Climate Neutal European Cities by 2030*. https://netzerocities.eu/.

[226] Sharing Cities Sweden. (n.d.). *Urban Sharing in Gothenburg*. https://static1.squarespace.com/static/59e86b55aeb625e2140eec1a/t/5bfb0fe64d7a9c8adb35d587/1543180277199/Urban+Sharing+in+Gothenburg+-+Snapshot.pdf.

[227] Viable Cities. (n.d.). *Together Towards Climate-neutral Cities*. https://en.viablecities.se.

[228] Sharing Cities Sweden. (n.d.). *Sharing Cities Sweden*. https://www.sharingcities.se.

civil society. The city has a sustainability-focused aim to change the composition of consumption to enhance resource efficiency, decrease per capita consumption, and consequently, reduce waste production. For example, the industrial area of Södra lvstranden is transformed from an industrial area into a living and sharing hub with sharing services. The city also supports local initiatives to spread the sharing culture, e.g., bike kitchens, which provide an accessible environment where people can learn to fix bicycles, share resources, and build a community. Other sharing services include the Smart Map of more than

Gothenburg Smart City Timeline

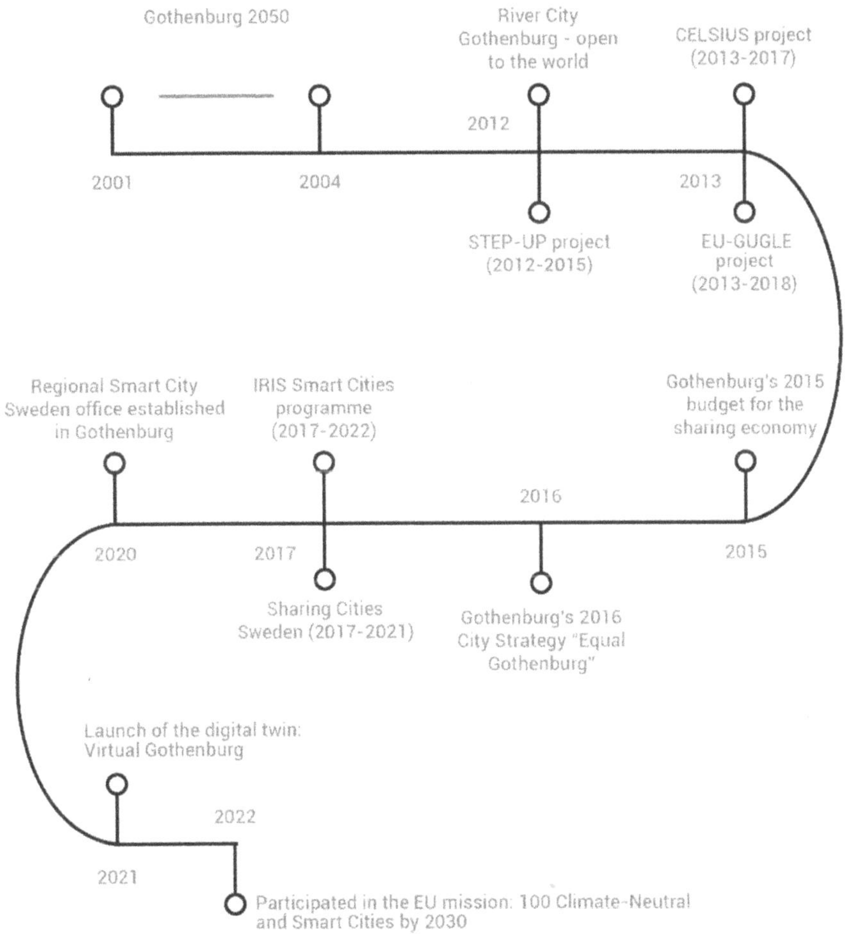

100 sharing activities, toy libraries, co-working and outdoor offices, and the West Sweden Tourist Board's Meet the Locals programme.[229]

3.3 Smart Tourism

Gothenburg is well experienced in hosting major international meetings and events. The city demonstrates the ideal combination of metropolitan advantages and a genuine small-town atmosphere. Meeting venues, accommodation options, dining establishments, and cultural attractions are situated close to each other.[230] Recognised among the most sustainable city destinations in the world, 97% of Gothenburg's public transportation is powered by renewable energy and 95% of its hotels have received eco-certification.[231]

The city has constantly performed well in international city rankings, notably being a six-year leader in the Global Destination Sustainability Index since 2016 and was named the winner of the 2020 European Capital of Smart Tourism.[232] In 2021, Lonely Planet announced Gothenburg as the Best Sustainable City Stay in the World.[233] Tourism in Gothenburg saw its first peak in 2015, bringing positive impacts to places such as the Landvetter Airport, Swedish Exhibition & Congress Centre, and Liseberg theme park, which enjoyed record levels of visitors.[234] Tourism made up 10.9% of Gothenburg's GDP in 2019, making it a significant contribution to the city's economy.[235] In 2021, there were a total of 3.9 million registered bed nights in Gothenburg, with 3.3 million spent by domestic

---

[229] *Ibid.*

[230] ICTTP 7. (n.d.). *Gothenburg*. https://www.icttp7.se/practical-info/gothenburg/.

[231] Bulter, A. (26 July 2022). *Op. cit.*

[232] Global Destination Sustainability. (2022). *Raising the Bar 2022 GDS Index Results Report* (p. 35). https://www.gds.earth/wp-content/uploads/Raising-the-Bar-2022-GDS-Index-Results-Report.pdf.

[233] Goteborg & Co. (2021). *Lonely Planet: Gothenburg is the World's Best Sustainable City to Visit 2021.* https://goteborgco.se/en/2020/11/lonely-planet-gothenburg-is-the-worlds-best-sustainable-city-to-visit-2021/.

[234] Göteborg Stad. (2015). *City of Gothenburg Annual Report* (p. 116). https://goteborg.se/wps/wcm/connect/57efa7c2-f9cb-46a8-b68b-b0ea4f977c43/City_of_Gothenburg_annual_report_2015.pdf?MOD=AJPERES.

[235] Göteborg & Co. (2019). *Annual and Sustainability Report 2019.* https://goteborgco.se/uploads/2020/03/ENG_GbgCo_Arsredovisning_Hallbarhetsredovisning_2019.pdf.

visitors and 587,000 by international travellers.[236] Tourism has become a substantial component in Gothenburg's economy and employment, making it pertinent for the city to constantly innovate and progress in the tourism and hospitality sector to continuously establish itself as a top global tourist destination.

Digital smart tourism in Gothenburg can be traced to 2018 when Göteborg and Co, a municipality-owned company published its Business Plan for Destination Gothenburg 2018–2020, which outlines Gothenburg's overarching goal of doubling its tourism numbers (to 9 million guest nights) by 2030 in a sustainable manner.[237] The five critical strategies include:

- **Taking collaboration to a new level**: This would be achieved through nurturing and enhancing pre-existing networks and stakeholder groups, creating new cooperation models, and experimenting with new methods of working and motivating young people and aspiring artists to join new networks.
- **Promoting Gothenburg globally**: Gothenburg's digital presence is promoted on the city's media channels and global platforms by showcasing the city's unique assets and events to pique international interest.
- **Creating innovation for the future**: Gothenburg aims to be a leader in digitisation and the use of new technology and be a facilitator of innovations.
- **Sharing a vibrant Gothenburg**: The city aims to share authentic, genuine, and engaging experiences about destination's unique features and support the emerging sharing economy.
- **Improving accessibility**: By supporting the city in its digital development and digital accessibility.

Tourism development is not a standalone activity. There is a complex relationship between tourism and climate action. Tourism contributes to and is impacted by climate change. Gothenburg is one of the signatories of The Glasgow Declaration on Climate Action in Tourism and is committed to accelerating climate action to reduce global tourism emissions

[236] TAK Research. (2022). *Greater Gothenburg — Accommodation and Travel Stats.* https://visitory.io/en/greater-gothenburg/.
[237] Göteborg & Co. (2018). *Op. cit.*

by at least half over the next decade and achieve net zero emissions before 2050.[238] Sustainability efforts made by Göteborg & Co are based on the 10 principles of the UN Global Compact, the UN World Tourism Organisation's definition of sustainable tourism and the Global Sustainable Development Goals. Göteborg & Co's mission aligns with the SDG Global Goals, focusing on three specific targets within those goals[239]:

- Target 8.9: Promote beneficial and sustainable tourism.
- Target 12.B: Develop and implement tools to monitor sustainable tourism.
- Target 17.16: Enhance the global partnership for sustainable development.

The COVID-19 pandemic has highlighted the importance of innovative and creative thinking in the rebuilding and reimagining of tourism. In early 2020, Göteborg & Co partnered with Group NAO to gather innovative and sustainable ideas for enhancing tourism experience in cities worldwide. The partnership resulted in a catalogue with 101 sustainable tourism ideas.[240] Another is the Destination Data Platform collaboration with Mastercard in response to the need for fast recovery after the pandemic.[241] In this partnership, Mastercard will deliver overall insights based on tourists' aggregated, anonymised network behaviour. The data can then give a sense of tourism demand — how tourists engage with the city's spaces and facilities (e.g., tourist attractions, food and dining, shopping, travel services, hotels, public transport, car rentals, and flights) before,

---

[238] Göteborg & Co AB. (2021). *Gothenburg Signs the Glasgow Declaration on Climate Action in Tourism.* https://via.tt.se/pressmeddelande/gothenburg-signs-the-glasgow-declaration-on-climate-action-in-tourism?publisherId=2567142&releaseId=3310001.

[239] Göteborg & Co. (2021). *Annual and Sustainability Report 2020.* https://goteborgco.se/uploads/2021/03/GbgCo_Arsredovisning_Hallbarhetsredovisning_2020_EN.pdf.

[240] Göteborg & Co. (n.d.b). *101 Ideas for Sustainable Tourism.* https://wewww.101sustainableideas.com/about/.

[241] Meeting Media Group. (2022). *Gothenburg: Where Sustainability is the Driving Force for Change*! http://meetingmediagroup.com/article/gothenburg-where-sustainability-is-the-driving-force-for-change.

during, and after their trip. According to Göteborg & Co, the collaboration will focus on four key areas[242]:

- Create and develop a common platform with anonymised and aggregated data.
- Deliver increased insight and knowledge about tourism in Gothenburg and the players in the hospitality industry.
- Inform decision-making about tourism supply and investment.
- Create the right conditions for research, development, and innovation in the hospitality industry.

The collaboration helps Gothenburg to create a common platform for the hospitality industry, leverage anonymised financial transaction data, and accelerate digital transformation while innovating, researching, and co-creating solutions across the tourist industry. Gothenburg is well connected to various partners and networks in the tourism industry for peer-to-peer learning in smart tourism including the following[243]:

- EU CVB Network is a convention bureau network of seven cities dedicated to sharing knowledge, experiences, and best destination practices.
- City Destinations Alliance (CityDNA) connects more than 130 metropolitan regions in 30 countries to promote tourism. In 2022, the annual CityDNA Tourist Information Centres Expert Meeting event was hosted in Gothenburg.[244]
- International Congress and Convention Association (ICCA) is the global community and knowledge hub for member cities to exchange experiences in the events industry. In 2020, Sweden hosted the ICCA Scandinavian Chapter Meeting in Gothenburg. The Chapter Meeting,

---

[242] Göteborg & Co. (2021). *Göteborg & Co Inleder Samarbete Med Mastercard för att utveckla Göteborgs besöksnäring.* https://goteborgco.se/2021/05/goteborg-co-inleder-samarbete-med-mastercard-for-att-utveckla-goteborgs-besoksnaring/.

[243] Göteborg & Co. (2023). *International Partnerships.* https://goteborgco.se/en/our-services/projects-and-collaboration/international-partnerships/.

[244] City Destinations Alliance. (2022). *TIC Expert Meeting 2022.* https://citydestinationsalliance.org/mpage/HomepageTIC2022.

which drew more than 80 attendees, had sustainability as its main topic. Topics discussed included the implications of changing travel patterns on the tourism industry and developing a standardised method for measuring the legacy and beyond-tourism effects of international congresses.[245]

3.4 Implementation and Impact of Gothenburg's Smart Tourism

Since the early 2010s, Gothenburg has been implementing a smart tourism ecosystem. The city has aimed to promote sustainable smart tourism through digital transformation. Focusing on enhancing tourism experiences and considering the needs of local communities, Gothenburg city authorities in partnership with civil society have invested resources to implement several smart innovations to sustain smart tourism systems. This section elaborates on the implementation and impact of two of these innovations: the Smart Map and Min Stad.

3.4.1 *The Smart Map*

Smarta Kartan (translated to English: Smart Map) is a website and digital map promoting the local sharing economy (Figure 3.1). The digital platform displays over 100 small-scale sharing projects in Gothenburg where locals and visitors may rent, borrow, share, exchange products, and contribute to services.[246] Smart Map aims to encourage community engagement and sharing over private ownership through public participation and sharing of activities, locations, and networks on the platform. Users would find amenities, services, and places shared by others on the map, such as bicycle parking, eateries, social spaces, shops, carpools, and various digital platforms.[247]

[245] Göteborg & Co. (2018). *ICCA Scandinavian Chapter — A Forum for a Growing Meeting City*. https://goteborgco.se/en/2018/04/icca-scandinavian-chapter-a-forum-for-a-growing-meeting-city/.

[246] Smart Map. (2022). *About the Smart Map. Smarta Kartan*. https://www.smartakartan.se/om-smarta-kartan/.

[247] Smart Map. (n.d.). *Smarta Kartan/Gothenburg*. https://www.smartakartan.se/en/gothenburg/.

Figure 3.1. Gothenburg sharing initiatives on the Smart Map website.
Source: Smart Map.[248]

The Smart Map was initiated and developed by Collaborative Economy Gothenburg (KEG); a civil society organisation founded in 2014. KEG is one of the most prominent actors in the local sharing economy.[249] In 2016, KEG formed an Idea-Based Public Partnership (IOP) with the City of Gothenburg's Democracy and Citizen Service administration, which regarded the sharing initiatives as an opportunity to promote a more sustainable lifestyle in Gothenburg. The partnership garnered €10,000 of public funding for the launch of Smart Map 1.0 in the same year.[250]

KEG and the City of Gothenburg Democracy and Citizen Service administration meet monthly to add new projects proposed by residents and ensure that the map's activities are relevant.[251] The activities presented on the map are determined by five main criteria[252]:

---

[248] *Ibid.*
[249] *Ibid.*
[250] Eurocities. (27 November 2019). *Rethink, Reuse and Repair*. https://eurocities.eu/stories/rethink-reuse-and-repair/.
[251] Eurocities. (n.d.a). *Smart Map: Rent, Share, Exchange, Borrow*. https://use.metropolis.org/case-studies/smart-map-rent-share-exchange-borrow#casestudydetail.
[252] Smarta Kartan. (n.d.). *Criterias*. https://www.smartakartan.se/en/criterias.

- **To facilitate access over ownership**: Rather than focusing on profiting, content posted should be consistent with the Smart Map's general theme of contributing to the sharing economy.
- **Open to everyone**: The criterion is intended to promote social inclusion and eradicate discrimination.
- **There are local users or participants**: The Smart Map would display events/to encourage local participation.
- **Small local initiatives**: The Smart Map would support and encourage small-scale services and projects.
- **Not widely known**: The Smart Map would display services and projects that are not well-known to the public.

In addition to featuring sustainable consumption initiatives, Smart Map encourages a broader meaning of sharing, such as companionship, meals, and experiences. The 'Meet the Locals' project is an example. Through Meet the Locals, West Sweden's visitors can connect with locals and experience the Swedish lifestyle and culture from a local perspective. For travellers, this might entail anything from enjoying their hosts' meals to hiking on their favourite trails or exploring off-the-beaten-path locations.

Within the first six months of its operation in 2016, a survey showed that 10,000 users had visited the Smart Map platform.[253] Many were hopeful about the platform, enabling them to live a more sustainable lifestyle. In 2019, Smart Map released an open-source tool, which allows other cities to replicate the Smart Map framework in their cities free of charge.[254] Funding and support of Sharing Cities Sweden contribute to the expansion of Smart Map from Gothenburg to include Malmö, Umeå, Stockholm, and Karlstad.[255] The City of Malmö and the Sjuhärad area have debuted local versions of the Smart Map. However, there are challenges to the Smart Map's sharing initiative. Barriers to sharing in

---

[253] Eurocities. (n.d.b). *Op. cit.*

[254] *Ibid.*

[255] Mccormick, K. & Leire, C. (Eds.). (2021). *Sharing Cities Sweden: Key Findings and Amplifying Impacts.* (Viable Cities Report Series; Vol. 2021, No. 8). Lund University.

Gothenburg include low community awareness of the Smart Map and citizen scepticism or an unfavourable attitude towards sharing.[256]

A survey conducted by the Consumer and Citizen Service in 2021 further highlighted varying opinions about sustainable consumption and the sharing economy.[257] Over 960 Gothenburg citizens participated in the survey. About 90% are optimistic about reducing their consumption over the next five years and 70% say they donate items they no longer intend to retain. The three most prevalent reasons why Gothenburg citizens use sharing economy include the following: it is good for the environment (63%), it saves money (51%), and it is practical (36%). However, a subset of residents (mainly older males from the lowest or highest household income families) have a negative perception and are opposed to using sharing services. Most are extremely negative about two activities: renting something from a private person and renting something to a private person. This negative attitude towards peer-to-peer (P2P) service whereby two individuals interact directly with each other without intermediation by a third party may be due to it being a relatively new service in the Swedish market.

What is perhaps more worrying is that after years of implementation, Gothenburg citizens did not seem to be familiar with several local sharing services. A small proportion, 15% of respondents, had never heard of any sharing services, while only 5% and 4% of respondents were aware of the Smart Map and Meet the Locals service, respectively. Respondents were also asked about trust and whether they agreed with the following statement: "It is possible to trust people in general". Of the 940 Gothenburg citizens who responded, 30% stated high trust, 50% stated medium trust, and 20% stated low trust. Even though Gothenburg is at the forefront of sharing economy development, the survey results reveal the knowledge-awareness gap and the necessity to increase the range of sharing services based on citizens' needs and to strengthen citizens' knowledge of the sharing service.

[256] Voytenko Palgan, Y., Mccormick, K., Leire, C., Lund, T., Öhrwall, E., Sandoff, A., & Williamsson, J. (2020). *Mobile Lab on Sharing in Gothenburg*. Lund University.

[257] Vårt Göteborg. (21 June 2021). *Göteborgarna positiva till delningsekonomi. Vårt Göteborg*. https://vartgoteborg.se/konsument/goteborgarna-positiva-till-delningsekonomi/.

Another challenge is financial resources. Budgetary shortfalls of the city government can serve as a potential hindrance to civil society organisations like KEG in their growth and expansion phase, especially during the COVID-19 pandemic.[258] Due to structural, political, economic, cultural, and sustainability concerns, governance of the sharing economy differs across cities.[259] Gothenburg has adopted a proactive approach to municipal regulation of the sharing economy, actively shaping and supporting the sharing environment through a variety of methods and roles. When faced with challenges during the innovation process, the city has discovered innovative methods of collaborating with sharing economy organisations by establishing the framework for a sustained partnership. The partnership activates collaboration among groups of local partners from the public sector, business, academia, and civil society and, in the process, augments financial resources and strengthens collaborative capital.[260] Another strategy is urban experimentation to create opportunities for urban change. Gothenburg has created testbeds under the Sharing Cities Sweden programme including urban living labs to learn from the social and technical innovations.

3.4.2 *Min Stad*

Gothenburg engages with many stakeholders through different operations and processes in the city. In 2012, the City Planning Administration developed a public tool called Min Stad (translated to English: My City) to aid urban development through participatory planning and citizen feedback. Min Stad is a digital bulletin board (available on both the web and as an application for mobile phones) where users can share their stories, needs,

---

[258] Persson, J. (16 June 2022). *Interview for Collaborative Economics Gothenburg* [Personal communication]; Lantto, K. (17 June 2022). *Interview with Development Leader Göteborgs Stad* [Personal communication]; Lund, T. (17 June 2022). *Interview with Sustainability Manager Göteborgs Stad* [Personal communication].

[259] Palgan, Y. V. (n.d.). *Gothenburg. Sharing & The City*. https://www.sharingandthecity.net/gothenburg.

[260] Lantto (17 June 2022). *Op. cit.*; Lund (17 June 2022). *Op. cit.*

Figure 3.2. Min Stad website.
Source: Min Stad.[261]

and interests.[262] The application is an interactive map where users can create suggestions within various categories, such as culture, recreation, cycling, and socialising (Figure 3.2). Min Stad aims to achieve the following through community participation:

- Create public meeting places.
- Involve public participation and engagement in city planning.
- Increase knowledge concerning city planning.
- Provide a platform for open discussion.

Both locals and visitors are welcome to use the application. Min Stad incorporates the use of Google and Facebook accounts to authenticate users who comment on the portal. Users have the option to share their posts on social media (Facebook and Twitter), leading to a wider distribution of each post and the promotion of the Min Stad portal. Posts from Min Stad are also uploaded into the internal operating system of the Urban Business Office. While posts are not answered by the City of Gothenburg,

[261] Min Stad. (n.d.). https://minstad.goteborg.se/.
[262] Klefbom, S. (2021). *Augmented Urban Values: Virtual Gothenburg as a Place for Citizen Dialogue and Shared Lived Experiences*. http://urn.kb.se/resolve?urn=urn:nbn:se:mau:diva-45934.

the inputs are used as inspiration and are referenced by officials in the Department of City Planning to aid their planning work.[263]

Citizens can send suggestions about urban development on Min Stad. Users can give direct feedback and report issues for any location within the map.[264] The application may be used by both individuals and the city to share stories or history about specific geographic regions within the city. When visitors or locals come across an intriguing place in the city, they can acquire information and examine historical images from the Gothenburg City Museum.[265] The city also utilises the application to convey information about the city's development or infrastructure projects, present and proposed. The platform includes interactive maps and visualisations to help individuals understand proposed plans and their possible influence on the city. They may also build small 3D models, such as buildings on the web portal. By turning the platform into an online hub containing all smart city plans: data describing ongoing plans, ongoing street work, planned events or documents, and information about political decisions, all geo-coded to a location or area, Min Stad acts as a communication channel to design more effective citizen and visitor engagement processes.[266] The platform has received international acclaim for allowing citizens and visitors to respond to and participate in local government plans. Awards include the Excellence Award at the 2012 Geospatial World Forum in Amsterdam and the Golden Link Award in 2013.[267]

Despite the initial acclaim, the platform has experienced several ongoing challenges. Between 2012 and 2020, the posting frequency on the platform decreased. The declining activity may be due to users' misperception of Min Stad as a communication platform to contact a specific city

---

[263] dialogGuiden. (2020). *My City Gothenburg*. https://dialogguiden.se/article/show/575.

[264] Lantto (17 June 2022). *Op. cit.*; Lund (17 June 2022). *Op. cit.*

[265] Vårt Göteborg. (2016). *Påverka Göteborgs utveckling med nya appen Min Stad*. https://vartgoteborg.se/tema-stadsutveckling/paverka-goteborgs-utveckling-med-nya-appen-min-stad/.

[266] Mangnus, A. C., Vervoort, J. M., Renger, W.-J., Nakic, V., Rebel, K. T., Driessen, P. P. J., & Hajer, M. (2022). Envisioning alternatives in pre-structured urban sustainability transformations: Too late to change the future? *Cities, 120*, 103466. https://doi.org/10.1016/j.cities.2021.103466.

[267] dialogGuiden. (2020). *Op. cit.*

department, expectation of some reply to their posting and resultant feeling that their efforts were in vain when no feedback was received.[268] The platform is also described as challenging to use since it covers the entire city's many projects. While it may be convenient to attain an overall perspective of urban development by including the entire city, it is sometimes more difficult to discover individual projects at a given location. In addition, users of Min Stad are mainly people interested in urban development, limiting its audience reach to a specific demographic. Also, there are insufficient resources to process all the citizen inputs to extract meaningful information.[269]

A 2019 study by the Chambers University of Technology suggested that Min Stad could prioritise visual content hierarchically to avoid a cluttered interface.[270] One strategy is to formulate a more sophisticated platform to allow citizens to post comments, amend the city's plan, and generate their own proposals in the form of maps. Another strategy is to harness Gothenburg's digital twin, Virtual Gothenburg, to become a future virtual destination.[271] The digital city could facilitate visualisation and simulation of various travel and tourism events and spaces.

3.5 Conclusion

Gothenburg's position on Sweden's scenic west coast has attracted visitors, domestically and globally. Gothenburg faces the challenge of over-tourism and has sought to become a sustainable tourist destination, a common response in many European tourist destination cities.

---

[268] Hellden, D. & Zhao, R. (2020). Motivation for Participation in City Development. A Case Study of Citizensourcing in Gothenburg, Sweden. [Master's thesis, Chalmers University of Technology]. https://odr.chalmers.se/bitstream/20.500.12380/302191/1/E2020_123.pdf.

[269] Klefbom (2021). *Op. cit.*

[270] Persson, I. & Tran, H. (2019). Designing for Dialogue. The Challenges of Facilitating Online Civil Dialogue for the Urban Planning Process. [Master's thesis, Chalmers University of Technology]. https://odr.chalmers.se/bitstream/20.500.12380/300513/1/CSE%2019-82%20ODR%20Persson%20Tran.pdf.

[271] Goteborg & Co. (2022). *Annual and Sustainability Report.* https://goteborgco.se/uploads/2022/03/2021_ENG_GoteborgCo_Arsredovisning_Hallbarhetsredovisning.pdf.

Gothenburg's tourism development has generated a greater demand for high-quality experiences and the city is embracing ICT to enhance local and visitor experiences and grow smart tourism.[272] Technology is a lever to enhance citizen and visitor experiences of the city. Gothenburg's approach aims to reshape visitor experience by providing smart tourism tools, including the Min Stad and Smart Map platforms to offer enhanced information, decision support, mobility, and ultimately, better travel experiences to both residents, tourists, planners, and service providers. Digitalisation processes do not replace but have created new demands for hybrid experiences and data-driven knowledge. As Gothenburg navigates its current tourism demands, the integration of digital tools refines its tourism landscape and at the same time, promotes sustainability, inclusivity, and community collaboration.

Projects like Smart Map also promote resource conservation through local sharing initiatives, while Min Stad facilitates citizen feedback and involvement in city planning and development. A key feature of these digital platforms is improving knowledge (e.g., knowing what and where services are available) and experience and product exchange (e.g., sharing experiences of places visited and renting instead of buying a particular tool/equipment). Gothenburg showcases how smart tourism growth strategies could be mobilised to advance wider sustainable development, potentially reducing its carbon emissions, and moving towards a zero-climate footprint while improving local livelihoods. The ambition is a part of the city's overarching sustainable development vision with long-term goals and strategic initiatives fostering social equity, environmental resilience, and eco-friendly solutions for a more inclusive and resilient future. The smart tourism development is not without its challenges, such as low awareness and financial constraints. Despite these challenges, the platforms have gained peer recognition for their innovative approaches.

Research is foundational to smart growth. It is not just a driver of technological advancement but also contributes to resolving roadblocks.

---

[272] Gretzel, U., Sigala, M., Xiang, Z., & Koo, C. (2015). Smart tourism: Foundations and developments. *Electronic Markets*, *25*(3), 179–188.

To address the challenges, studies have been conducted to better understand and evaluate project outcomes. The suggestion is to make Min Stad more attractive and easier to navigate, e.g., enhancing Min Stad's interface with hierarchical visual content and leveraging other ongoing smart projects, like the Virtual Gothenburg digital twin as a future virtual destination for immersive tourism experiences. Collaboration and partnerships are central to the development effort. Innovation challenges of the Smart Map, e.g., limited resources, could be mitigated by fostering sustained partnerships with sharing economy organisations, pooling resources from diverse sectors, and fostering local collaborative capital.

CHAPTER 4

HANGZHOU: SMART RETAIL

4.1 City Overview

Location[273]	Hangzhou, Capital City of Zhejiang Province, China
Land Size[274]	16,850 km^2
Climate	Subtropical monsoon
Geography (topology/landscape)[275]	Mountainous terrain, plains, and water surfaces
Environmental Risks (if any)[276]	Water pollution of organic and inorganic substances, eutrophication
Population[277]	11,936,000 (2020)
GDP[278]	US$160 billion (2020)

[273] Hangzhou, China. (n.d.). *Op. cit.*

[274] *Ibid.*

[275] *Ibid.*

[276] Zhu, G., Noman, M. A., Narale, D. D., Feng, W., Pujari, L., & Sun, J. (2020). Evaluation of ecosystem health and potential human health hazards in the Hangzhou Bay and Qiantang Estuary region through multiple assessment approaches. *Environmental Pollution, 264*, 114791. https://doi.org/10.1016/j.envpol.2020.114791.

[277] Hangzhou Bureau of Statistics (2021). *Op. cit.*

[278] Hangzhou Municipal Statistics Bureau — HMSB (2021), Second land survey.

GDP Per Capita[279]	US$19,000 (2020)
Key Urban Issues of Smart City[280]	Industrial overcapacity, unsustainable manufacturing practices, outdated business practices

Hangzhou is situated in south-eastern China and classified as a second-tier city.[281] It has a land size of 16,850 km^2 and a population of 11.93 million in 2020.[282] This is an increase of 3.2 million residents since the last census in 2010.[283] The City of Hangzhou has an urbanisation rate of 83.3% with a population of 9.94 million in urban areas. The city's average density is 708 persons per square kilometre.[284] Its landscape is predominantly made up of hilly or mountainous terrain (65.6%), plains (26.4%), and water surfaces (8%).[285] Mountainous terrain dominates the west, middle, and south of the city, while the north-east area is covered with plains and water surfaces.[286] Hangzhou has a subtropical monsoon climate with distinct seasons. 2020 registered an annual average temperature of 18.3°C, with a total precipitation of 1,664 mm and 1,619 hours of sunshine; the highest recorded temperature was 39°C on August 15 in downtown Hangzhou,

---

[279] *Ibid.*

[280] Shanghai ATO & USDA. (2018). Emerging city market report — Hangzhou (Market Development Reports No. CH178042; p. 30). USDA Foreign Agricultural Service.

[281] Shanghai ATO & USDA. (2018). *Op. cit.*; Hernández, M. (2016). China City Tiers. *South China Morning Post.* https://multimedia.scmp.com/2016/cities/; Cities in China are categorized into four tiers (Tiers 1–4, Tier 1 indicating cities with the largest GDP and population size. This also determines their structure of governance. Tier 1 cities are directly controlled by the central government, while the rest are divided into provinces, sub-provinces, prefectures, and counties.)

[282] Hangzhou Bureau of Statistics. (2021). *China's 7th Population Census, Municipal Statistical Yearbook of Hangzhou.*

[283] Hangzhou Bureau of Statistics. (2010). *China's 6th Population Census, Municipal Statistical Yearbook of Hangzhou.*

[284] Hangzhou Bureau of Statistics (2021). *Op. cit.*

[285] Hangzhou, China. (n.d.). *Hangzhou China About Hangzhou.* http://eng.hangzhou.gov.cn/col/col1450889/index.html.

[286] *Ibid.*

while the lowest temperature was –7.3°C on December 31 in the district of Lin'an.[287]

Hangzhou is one of China's leading cities in industrialisation and information integration, accelerating its smart city innovations. Hangzhou's GDP per capita was recorded at US$20,000 in 2020, providing employment for a total of 7.5 million people, with the primary, secondary, and tertiary industries accounting for 7%, 34.9%, and 58%, respectively.[288] The city's total GDP stood at 1.2 trillion yuan in 2020, with an annual growth rate of 3.9%. This is 1.6% higher than the national average and 0.3% higher than the provincial average.[289] Hangzhou's economic structure is split into three main industries — primary (agriculture), secondary (construction and manufacturing), and tertiary (service sectors) industries. They contribute 32.6 billion yuan, 482.1 billion yuan, and 1.1 trillion yuan respectively to the city's GDP.[290] The digital sector constitutes a large portion of Hangzhou's tertiary industry; 26.6% of local GDP in 2020 was from core digital industries comprising e-commerce (193.3 billion yuan), cloud computing and big data (138.9 billion yuan), digital contents (311.3 billion yuan), software and information services (344.1 billion yuan), and e-information products manufacturing (109 billion yuan).[291] E-commerce in Hangzhou contributes 21.7% of the city's total exports, amounting to 75.7 billion yuan worth of total export value (excluding provincial-level companies).[292]

Hangzhou is a subprovincial city under the province of Zhejiang. As a second-tier city, Hangzhou is managed by a governor appointed by the Chinese Communist Party (CCP)[293] and guided by top-down policies issued by the Central Committee of the Chinese Communist Party (CPC) through the State Council.[294] The Provincial Government of Hangzhou oversees the

---

[287] *Ibid.*

[288] *Ibid.*

[289] *Ibid.*

[290] *Ibid.*

[291] *Ibid.*

[292] *Ibid.*

[293] Hernández (2016.). *Op. cit.*

[294] Ministry of Foreign Affairs China. (n.d.). *Administrative Division System.* https://www.fmprc.gov.cn/mfa_eng/ljzg_665465/zgjk_665467/3572_665469/t1140993.shtml.

running of all aspects of the city based on assigned divisions. The CCP Hangzhou Committee members make up the Hangzhou municipal government, with Liu Xin as the mayor of Hangzhou.[295] The government sets the trajectory of private companies through state policies, controls the prices of key commodities and financial policies, and allocates state financial investments and energy resources. This also includes state control over major industries for manufacturing and strict monitoring of telecommunication companies.[296] The City of Hangzhou has ten urban districts, one county-level city, and two counties. There are a total of 191 towns (subdistricts), which are further administratively broken down into townships, towns, subdistricts, neighbourhood committees, and administrative villages.[297]

4.2 Smart City Development: Economic Sustainability

Hangzhou has developed a smart city vision, focusing on the development of e-platforms and artificial intelligence (AI) to improve its public service and private commerce.[298] Newer initiatives also use big data and cloud computing to efficiently manage civic services and citizen welfare.[299] The Zhejiang province has a unique economy, which relies on private firms rather than state-owned enterprises due to its historical entrepreneurial traditions and decentralised governance.[300] Since industrialisation, the local industry comprised low-tech industrial clusters that specialised in various aspects of manufacturing. However, this economic model soon

---

[295] Hangzhou China. (n.d.). *Op. cit.*

[296] Encyclopedia Britannica. (n.d.). *China — The Role of the Government*. https://www.britannica.com/place/China; The State Council. (n.d.). *The State Council of the People's Republic of China*. http://english.www.gov.cn/.

[297] Hangzhou, China (n.d.). *Op. cit.*

[298] Atha, K., Callahan, J., Chen, J., Drun, J., Green, K., Lafferty, B., McReynolds, J., Mulvenon, J., Rosen, B., & Walz (2020). *Op. cit.*; The State Council. (2020). *Free Trade Zones Rise to 21 with New Approvals*. http://english.www.gov.cn/policies/policywatch/202009/22/content_WS5f693628c6d0f7257693c63a.html.

[299] Hangzhou.gov.cn. (2021). *City Brain*. http://eng.hangzhou.gov.cn/art/2021/5/31/art_1229498047_58875531.html.

[300] Wei, Y. D. & Ye, X. (2004). Regional inequality in China: A case study of Zhejiang Province. *Tijdschrift Voor Economische en Sociale Geografie, 95*(1), 44–60.

became unsustainable, especially with the region's rapid modernisation and transition into high-tech industries. Older industries were running at overcapacity and were unable to sustain low-end manufacturing due to the scarcity of labour, capital, and other resources.[301] This resulted in the need for governmental intervention when hit by the global financial crisis in 2008.[302] The government provides financial investments to encourage market entry and expansion of Smart Cities technology.[303] Guiding policies are issued by the Central Committee of the CPC through the State Council, aligning with regional development patterns.[304] Resultingly, local-scale plans for smart developments have been developed, detailing Hangzhou's roadmap for industrial revitalisation:

- **Hangzhou Master Plan for the Development of Ten Major Industries (2011–2015)**[305]: This master plan fostered scientific and innovative development in urban sectors. These included ICT, energy, and transportation to reflect the city's vision for the development of technological and economic innovation to drive economic growth. Targeted locations or "zones" were allocated for spatial clustering of technology-based enterprises, local financial innovation, and professional training and skills development. This enabled local economic and industrial structuring and the development of new and high-tech industries.
- **Smart Hangzhou Construction Master Plan (2012–2015)**[306]: This master plan was released after the national-level Future Sci-Tech City was approved in 2011. Hangzhou's municipal government has

[301] *Ibid.*

[302] Gong, J., Boelens, L., & Hua, C. (2020). How can China achieve economic transition with the featured town concept? A case study of hangzhou from an evolutionary perspective. *Urban Policy and Research, 38*(3), 249–261.

[303] Atha *et al.* (2020). *Op. cit.*

[304] Guofa. (2018). *Op. cit.*

[305] Argyriou, I. (2016). Planning the smart city in China: Key policy issues and the case of dream town in the city of Hangzhou. In *Proceedings of the 25th International Conference Companion on World Wide Web.*

[306] Hangzhou.gov.cn (31 January 2018). 2017年杭州经济形势发布新闻稿 http://www.hangzhou.gov.cn/art/2018/1/31/art_805866_15433455.html; Shanghai ATO & USDA. (2018). *Op. cit.*

transformed the city's industrial structure since 2013 by increasing its R&D spending. By investing in a variety of high-tech manufacturing industries, such as new energy vehicles, biopharmaceuticals, robotics, and aerospace components, Hangzhou has propelled the agenda of creating an entrepreneurship centre of "Internet Plus" with global reach.

- **Smart Economy Project (2014)**[307]: Hangzhou launched the Smart Economy Project to promote the fourth industrial transformation and upgrading to make breakthroughs in new areas of the digital economy, such as AI, Internet of Things, and virtual reality.
- **Construction of Innovation Platforms (2018–present)**[308]: Hangzhou plans to become the nation's foremost city in terms of the digital economy, focusing on three major development paths — industrialisation of digital resources, digitalisation of industries, and digitalisation of urban areas. Hangzhou will focus on the innovation and application of digital technologies and increase support for major platforms and projects. Construction of innovation platforms, such as Zhejiang Laboratory, Westlake University, and Alibaba Damo Academy, has accelerated, making the city an ideal place for tech-based entrepreneurship and innovation.

4.2.1. *The Alibaba Company*[309]

Despite clear top-down plans for revitalisation, Hangzhou still maintains its unique economy through strong private sector backing and initiative

[307] Zhu, W. & Chen, J. (2022). The spatial analysis of digital economy and urban development: A case study in Hangzhou, China. *Cities, 123,* 103563. https://doi.org/10.1016/j.cities.2022.103563.

[308] Wang (22 October 2018). Hangzhou is Building the Top City of China's Digital Economy. *Bloomberg*. https://www.bloomberg.com/press-releases/2018-10-22/hangzhou-is-building-the-top-city-of-china-s-digital-economy; 陈子琰. (19 September 2019). Hangzhou at Forefront of Digital Economy. *China Daily*. https://www.chinadaily.com.cn/a/201909/17/WS5d8075a4a310cf3e3556bed1.html.

[309] Alibaba Group. (2021). 2021 Fiscal Year Annual Report (Annual Report No. 9988).; Liu, X. (6 June 2016). The Success of Alibaba Has Turned Hangzhou into an Enviable e-Commerce Hub. *Global Times*. https://www.globaltimes.cn/content/987201.shtml;

Hangzhou Smart City Timeline

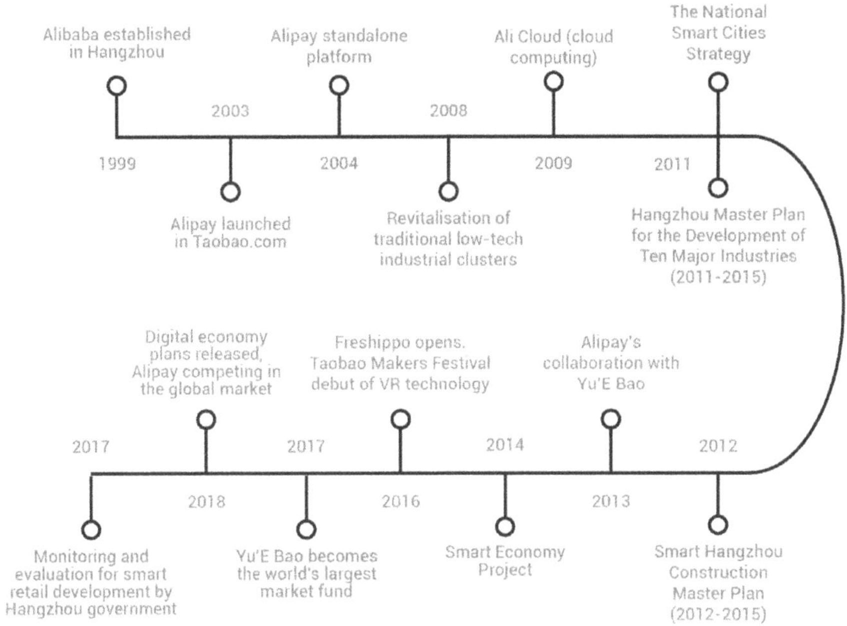

for the city's innovations and development. A key player in the city is Alibaba, established in Hangzhou by Jack Ma in 1999. The company has a significant presence and impact on the e-commerce industry of Hangzhou. It has since grown into a leading online and mobile commerce conglomerate, helping shape internet industries in the city and internationally boost the overall development of e-commerce. Alibaba was first established as a wholesale online marketplace, which connected Chinese manufacturers with overseas buyers. Alibaba subsequently expanded to create a portfolio of platforms, covering every aspect of e-commerce. Taobao.com supports customers-to-customers (C2C) interface, Tmall.com, Juhuasuan.com connects businesses to customers (B2C), while

---

Wu, X. & Gereffi, G. (2018). Chapter 13 Amazon and Alibaba: Internet Governance, Business Models, and Internationalization Strategies. In R. van Tulder, A. Verbeke, & L. Piscitello (Eds.), *Progress in International Business Research* (pp. 327–356). Emerald Publishing Limited.

Alibaba.com, their first platform, connected businesses to businesses for larger wholesale purchases (B2B). Alibaba also diversified into other industries by setting up a travel agency (Alitrip), a logistics network (Cainiao), fintech and online payment services (Alipay), and cloud computing to optimise logistic and financial processes (Ali Cloud). In 2021, Alibaba has grown significantly with an annual revenue reported at 717,289 million yuan, up from 101,143 million yuan in 2016. Their e-commerce earnings in China total 473.7 billion yuan, accounting for 66% of their annual revenue. At the same time, their international sale stands at an annual sum of 34.5 billion yuan.

4.3. Smart Retail and Marketplace

Hangzhou's traditional retail industry was on the decline before 2016, particularly due to the emerging presence of e-commerce. The decline demonstrated the need for additional online retail options or new innovations in the retail industry.[310] A pivotal moment in Hangzhou's retail transition was during the 2016 Yunqi Town conference where Jack Ma gave a speech, highlighting the importance of e-commerce in the future economy and the transition into integrated platforms that host new retail, new manufacturing, fintech, new technologies, and new resources (data).[311] This conference paved the way for traditional and online retail markets to co-exist, illustrating a roadmap for the future of retail. It simultaneously functioned as a call to action for the community, gaining the attention of local government. [312]

Hangzhou's well-developed economy and higher GDP are the key driving factors of the city's success in establishing a bottom-up digitally based economy through smart retail and marketplaces.[313] Hangzhou is home to highly skilled residents with strong technical capabilities[314] in

[310] 汪秘书长 (Secretary-General Wang) (13 July 2022). *Op. cit.*

[311] AliCloud — Cloud Seal. (13 October 2016). 马云：“电子商务”将消失 新五通一平将引领未来-阿里云开发者社区. https://developer.aliyun.com/article/61583.

[312] 汪秘书长 (13 July 2022). *Op. cit.*

[313] *Ibid.*

[314] Argyriou, I. (2019). The smart city of Hangzhou, China: The case of dream town internet village. In L. Anthopoulos (Eds.), *Smart City Emergence* (pp. 195–218). Elsevier.

combination with their culture of entrepreneurship, which are often self-supported and implemented before the government takes notice and provides additional support.[315] This is notably due to Alibaba's headquarters and other major internet companies, which are significant in Hangzhou and have contributed greatly to the city's digital infrastructure.[316] Due to the cooperation and collaboration of traditional industries, their IoT development was accelerated. Apart from citizen-based efforts, the Hangzhou government also provides continuous policy support and guidance, which influence the capital investments the industry receives. This cultivates an ecosystem of support for the new retail economy by attracting young talents through backed investments and good career prospects. Specifically, the government is adopting new marketing approaches to keep up with the times while attracting a younger demographic through live-streaming services (on social media using trending topics or online celebrities).[317]

Hangzhou is now one of the leading cities in China's smart city development, looking at four key areas of urban development related to improving economic prosperity[318]:

- Digital infrastructure development
  - Integrating physical and digital data using sensors, IoT, and big data.
  - Pushing towards public engagement.
  - Improving digital and online security and risk management.
- Governance and urban sensing transformation through the City Brain platform
  - Data-driven innovation adopting an integrated approach to digital governance.
  - Enabling data integration across different domains, industries, and government departments.
- Digital governance to facilitate better civic service provision
  - Online platforms to host citizen feedback.
  - Citizen dialogue influencing digital governance policies.

---

[315] 汪秘书长 (13 July 2022). *Op. cit.*

[316] *Ibid.*

[317] *Ibid.*

[318] *Ibid.*

- Integrated development of core industries
  - Digital economy is to be integrated with traditional core industries in Zhejiang province.

Hangzhou government actively promotes its smart city programme through urban development policies to support the agglomeration of digital economy and local government's role in entrepreneurial urban governance.[319] To encourage enterprises to set up in Hangzhou and its development zones, local government provides tax breaks for high-tech firms, with additional incentives for internet start-ups, housing subsidies for entrepreneurs, and events to promote private and foreign investment to support start-ups.[320]

The Hangzhou Future Sci-Tech City (FSTC) developed in 2015 is an innovation and venture complex for global technology entrepreneurship. FSTC's objective is the resettlement of high-quality overseas talents and young professionals in strategic economic areas, such as the information economy, biomedicine, new energy, and financial services.[321] FSTC was first initiated in 2010 by the Chinese Central Organization Department and the state-owned Assets Supervision and Administration Commission of the State Council as a key implementation mechanism of China's National Strategy on Talents. It is part of Zhejiang's 100 Featured Town project.[322] Zhejiang's regional government initiated the Zhejiang Featured Towns projects, and the first towns were unveiled in 2015, comprising 37 specialised towns, 9 of which were in Hangzhou.[323] Zhejiang Featured Towns spatially catered to the growth of Hangzhou's high-tech information-based industries.

---

[319] Qian, Z. (2011). Building Hangzhou's new city center: Mega project development and entrepreneurial urban governance in China. *Asian Geographer, 28*(1), 3–19.

[320] Zhang, Z. (6 April 2021). Investing in Hangzhou, Zhejiang Province: China City Spotlight. *China Briefing News*. https://www.china-briefing.com/news/investing-in-hangzhou-zhejiang-province-china-city-spotlight/.

[321] Argyriou (2019). *Op. cit.*

[322] Bay Area Council. (n.d.). *Hangzhou Office: Future Sci-tech City*. https://www.bayareacouncil.org/policy/hangzhou-office-future-sci-tech-city/.

[323] News.Sina. (25 June 2015). 未来三年，浙江要建100个特色小镇. http://news.sina.com.cn/c/2015-06-25/052031985193.shtml; Gong *et al.* (2020). *Op. cit.*

The urban agglomeration strategy was aimed at accelerating innovation in the existing specialised fields, upgrading the industrial value change, and promoting a sustainable integrated rural and urban development model.[324] Many leading IT-sector firms have emerged through the provision of urban agglomeration conditions.[325] Subsidiary companies of Alibaba Group Holding Limited have since worked with provincial and local governments to further develop Dream Town and Yunqi Town[326]:

- **Dream Town**: FTSC's first development in Hangzhou was "Dream Town", a key incubation space for internet start-ups in the FSTC.[327] Dream Town spans 3.5 km^2 and integrates historical and future urban elements, protects existing water channels and farmland, and creates a space for car-free social activities.[328] Dream Town serves to finance and nurture an agglomeration of start-ups through the promotion of local innovation, contributing to new innovations for long-term sustainability.[329] The spatial proximity consolidates new innovations, attracting university graduates from the surrounding province in the areas of internet and finance.[330] Zhejiang province planned and funded the Dream Town urban infrastructure, while Alibaba, its subsidiaries, and spin-offs were responsible for the innovations and business expansion of this initiative.[331] Innovation and talent creativity drive growth and place-based competitiveness.[332] At present, Dream Town hosts nearly 300 start-ups with the goal of reaching 10,000 and attracting over 300 billion RMB (equivalent to over US$45 billion) of investment within the first three years of its operation. The investment target was achieved by 2016, within its first three years of operation.[333]

---

[324] *Ibid.*

[325] *Ibid.*

[326] Shanghai ATO & USDA (2018). *Op. cit.*; Gong *et al.* (2020). *Op. cit.*

[327] Argyriou (2019). *Op. cit.*

[328] *Ibid.*

[329] Gong *et al.* (2020). *Op. cit.*

[330] *Ibid.*

[331] *Ibid.*

[332] Argyriou (2019). *Op. cit.*

[333] *Ibid.*

- **Yunqi Town**: This is another urban agglomeration zone initiated by Alibaba with 30 other private firms. This would serve as the hub of cloud computing, hosting data centres that support on-demand access to computing resources, applications, servers, data storage, development tools, networking tools, and more.[334] Many "cloud" enterprises such as AliCloud, DT Dream, Zhengcaiyun, Xinhua Zhiyun, and Watong Cloud have settled in Yunqi Town. To draw recognition for innovations from Yunqi Town, Alibaba has been organising cloud computing conferences, attracting over ten thousand attendees in the years leading up to 2020.[335] Yunqi Town aims to generate 1.5 billion yuan (US$214 million) in general financial revenues by 2022 and 60 billion yuan in digital economy output.[336] This will be achieved through further expansion with new industrial parks of 0.53 km^2 to the west, focusing on research of aerospace information technology and a new industrial area to the south, spanning 4.9 km^2 for advanced manufacturing. The southern town also plans to attract high-level cloud computing talents and entrepreneurial teams by building a demonstration base and providing support for independent innovation in the cloud computing industry. These developments are supported through favourable policies that boost industrial growth and attract skilled workers.[337]

Supply-side structural reform is another key aspect of economic development, with Hangzhou acting as one of China's 21 Pilot Free Trade Zones.[338] It pioneers China's new model of bonded import and retail

[334] IBM (16 September 2021). *What is Cloud Computing?* https://www.ibm.com/cloud/learn/cloud-computing.

[335] Gong *et al.* (2020). *Op. cit.*

[336] E-Hangzhou.gov. (27 December 2019). Yunqi Town Aims to Be China's No. 1 Digital Town. *China Daily*. http://zhejiang.chinadaily.com.cn/2019-12/27/c_445403.htm.

[337] Hangzhou Xihe District (27 May 2020). *To Build the First Digital Economy Town in China, Yunqi Town Has Embarked on a New Journey of Version 2.0 Construction.* http://en.hzxh.gov.cn/art/2020/5/27/art_1208248_46603460.html.

[338] The State Council (22 September 2020). *Op. cit.*

processing,[339] with a new e-commerce demonstration centre being built.[340] The city also pilots a development zone for China's new generation of AI and financial technology (fintech) for a national-scale programme.[341] These programmes and innovations support the growth of local and global cross-border e-commerce, making Hangzhou the focal point of e-commerce business in China since 2015.[342]

4.4 Implementation and Impact of Smart Retail and Marketplace

This section elaborates on the implementation and impact of two of these innovations: hybrid retail and e-wallets.

4.4.1 *Hybrid Retail: The New Retail Model*

The New Retail model features network effects generated by combining digital capabilities and physical assets. This model experiments with the use of analytics and retail-as-a-service business. Powered using data analytics to inform consumer patterns and support back-end logistics, this approach demonstrates new innovations in supply chain strategies and technologies.[343] Alibaba first proposed this model where the company provides online platforms and logistics for business owners and customers to connect while simultaneously generating data and revenue through e-platform usage.[344] Alibaba's digital smart retail and marketplace services

[339] Hangzhou, China (n.d.). *Op. cit.*

[340] E-Hangzhou.gov (24 September 2020). *Hangzhou area of Zhejiang FTZ unveiled.* http://www.ehangzhou.gov.cn/2020-09/24/c_274883.htm.

[341] Hangzhou, China (n.d.). *Op. cit.*

[342] Wright, K. (21 July 2015). China Vamps up Hangzhou's e-Commerce Model. *China Briefing News.* https://www.china-briefing.com/news/china-vamps-hangzhous-e-commerce-model/.

[343] Achim, A.-L. (10 March 2019). How Jack Ma's 'New Retail' Model Puts China Ahead of the Game. *South China Morning Post.* https://www.scmp.com/magazines/style/news-trends/article/2188823/alibaba-gucci-how-jack-mas-new-retail-model-puts-china.

[344] Wu & Gereffi (2018). *Op. cit.*

are supported by Ali Cloud, created in 2009. Ali Cloud — Alibaba's cloud computing sector — stores and analyses data generated by e-commerce. Data are processed to extract features from a customer's recorded history of shopping behaviour.[345] These data help platforms recommend relevant products to individual customers based on their recorded preferences to increase product sales. Sellers are also able to bid for their listings to show up on e-platforms with a higher ranking among similar products when customers input certain keywords.[346]

Freshippo (originally known as Herma Fresh) is Alibaba's venture into establishing a hybrid network of e-platforms and assets.[347] The first physical store opened in 2016 and presented itself as a customer experience while functioning as a consumption and logistics centre (Figure 4.1).[348] Customers are given the option to shop traditionally in-store, pre-order their items online to collect in-store, or have products delivered to their doorstep. The payment and logistics are supported by the Freshippo or Taobao app, integrated with Alipay e-wallet. Freshippo plans to perform as market centres in local neighbourhoods by offering wider selections and higher quality merchandise with convenient logistics through the following[349]:

Figure 4.1. Freshippo hybrid retail.

Source: Authors.

[345] Ali Cloud. (n.d.). *Machine Learning Platform: Data Mining & Analysis*. https://www.alibabacloud.com/product/machine-learning.

[346] Wu & Gereffi (2018). *Op. cit.*

[347] *Ibid.*

[348] Bird, J. (12 November 2018). 11 Insights from 11/11, Alibaba's Singles' Day and the World's No. 1 Shopping Day. *Forbes*. https://www.forbes.com/sites/jonbird1/2018/11/12/11-insights-from-1111-the-worlds-1-shopping-day/.

[349] Bird, J. (18 November 2018). Alibaba's 'New Retail' Revolution: What is It, and is It Genuinely New? *Forbes*. https://www.forbes.com/sites/jonbird1/2018/11/18/alibabas-new-retail-revolution-what-is-it-and-is-it-genuinely-new/; Hou, Y. (2018). Herma — The Pathfinder of Alibaba's New Retail [Investor's Report]. Herma, Alibaba.

- Providing customers with quality assurance through options to return and refund items.
- Ethically sourcing their projects while making their origin and manufacturing chain transparent to customers.
- Giving customers the option to view the item's origin and journey from farm to shelf through their app.
- Letting customers access product details, ratings, and reviews just like they would online to compare different products.

Freshippo has driven business innovation in Hangzhou following the success of its hybrid model, influencing other major competitors and local businesses to implement smart retail capabilities. Examples include JD.com's smart mirror in shops that display information and details of each product, such as production area, weight, and nutritional facts.[350] Unstaffed retail has extended to restaurant concepts in Hangzhou, embracing digital transformation, allowing customers to avoid queues, and reducing the labour of staff. Smartphone applications allow customers to order and pay for food, while the restaurant offers a 24-hour space with automated vending machines to sell frozen food and drinks.[351]

Unstaffed retail shops require three major components for their operations[352] — identifying the consumer, service support platforms, and unstaffed payment. Alibaba achieves this by using Ali Cloud and its existing Taobao app to identify and record data on its consumers. It then utilises the Freshippo app and back-end platforms to keep track of the logistics, which manage and stock their shops. Lastly, the customer checks out using their Alipay system e-wallet.[353] Marketing is targeted to individual customer profiles, while supply chains are optimised based on real-time data. Alibaba's existing e-commerce logistics system is also

---

[350] Suchit, L.-N. (20 August 2018). Chinese e-Tailing Made Simple. *Bangkok Post*. https://www.bangkokpost.com/tech/1524958/chinese-e-tailing-made-simple.

[351] *Ibid.*

[352] Xu, J., Hu, Z., Zou, Z., Zou, J., Hu, X., Liu, L., & Zheng, L. (2020). Design of smart unstaffed retail shop based on IoT and artificial intelligence. *IEEE Access, 8*, 147728–147737.

[353] Hou (2018). *Op. cit.*

applied to ensure intelligent fulfilment of orders and rapid delivery within a span of 30 minutes in designated neighbourhoods.[354]

Following the uptake of smart retail in Hangzhou, the government has turned its focus on scaling up these innovations. As of 2022, Hangzhou government has established the following indicators to determine the success of hybrid retail shops[355]:

- Eighty percent of all new retail shops should have a digital price tag/QR code or digital payment methods.
- Implement 24-hour unmanned shops.
- Leading F&B enterprises should have online booking and service functions.
- Five streets in Hangzhou should be allocated for new retail (urban agglomeration).
- Thirty plus companies are being nurtured in the new retail industries.
- Community experience should be enhanced through new retail.
- Fresh produce collection lockers should be set up within 3 km of every neighbourhood or within 30 mins of walking distance.

Innovations, such as Freshippo, demonstrate the potential of retail automation, which reduces the amount of human labour required in store for operations. Workers can be reorganised to fulfil external logistics and shipments to online buyers.[356] This innovation solves two main problems in existing physical retail — customer identification and the logistics of stocktaking.[357] Consumers can be identified based on smartphone information, where biometric data can be used for identity authentication.[358] These sets of data contain customer's shopping habits, family background, financial status, and other interests.[359] Since Freshippo collects real-time

[354] *Ibid.*
[355] 汪秘书长 (13 July 2022). *Op. cit.*
[356] Bird (18 November 2018). *Op. cit.*
[357] Xu *et al.* (2020). *Op. cit.*
[358] *Ibid.*
[359] Hou (2018). *Op. cit.*

data on customer purchases and habits, it also digitally solves the logistics issue of item identification and availability.[360] Merchants can improve the sustainability of the supermarket model by stocking and predicting to some extent the right amount of fresh product that should be on the shelves in each store.[361]

Some of the main challenges faced during the transition to digital smart retail and marketplaces include the following[362]:

- Mindset shifts to embrace new technology.
- Financing new hard and soft infrastructure.
- Lack of available talent to manage new technology.
- Financial security management.
- Issue of commercial monopoly.

Financing the new infrastructure is the most critical challenge to overcome for the new retail transition. The retail industry has a small profit margin regardless of online or offline retail. Consequently, it is often difficult for small businesses to cover the start-up or transition cost for new infrastructure.[363] Freshippo relied greatly on Alibaba's initial investment to set up its large high-tech compound. The operation cost for online retail is also quite large due to the cost of running a high-capacity online server, depending on the number of online visitors. Hangzhou's government investment schemes for new retail play a very important role in funding and support to facilitate this retail transition.[364]

Hangzhou's vision and direction towards hybrid retail address challenges at the level of development and implementation but lack consideration for societal understanding and adaptation.[365] Current scaling up of hybrid retail in the city is being hindered by traditional enterprise mindsets, especially with small businesses that are apprehensive over the shift

---

[360] Xu *et al.* (2020). *Op. cit.*

[361] Bird (18 November 2018). *Op. cit.*; Xu *et al.* (2020). *Op. cit.*

[362] *Ibid.*

[363] 汪秘书长 (13 July 2022). *Op. cit.*

[364] *Ibid.*

[365] *Ibid.*

into an online-to-offline (O2O) model.[366] There are often common misunderstandings over the definition of hybrid retail which has been caused by the overly quick and varied development of new innovations (augmented reality, virtual reality, and the metaverse).[367] Some misconceptions still exist regarding online to offline retail and vice versa. Smart hybrid e-commerce requires an understanding of both the online and offline models, each comprising different elements of business management. For example, online platforms require social media engagement, business data analytics, and other user engagement factors to be successful. Meanwhile, offline retail would require large upfront capital and the management of physical infrastructure to support the logistics, supply chain and quality control. For hybrid retail, online and offline stores should be seen as part of a matrix that works as different components of the larger thing rather than separate practices.[368] In response to this challenge, the local government has provided comprehensive policy support and guidance for local businesses such as the "new retail high-quality development model" policy guideline which was published in June 2022.[369]

4.4.2 *E-wallets: Digital Payment and Investments*

E-wallets are predominantly smartphone-based platform applications that allow customers to make payments digitally. This platform has become a key part of everyday retail in China.[370] There are two major competitors in China — Alipay and WeChat Pay — that hold 92% of this industry's market share.[371] Consumers utilise this platform for payments in several ways[372]:

---

[366] *Ibid.*

[367] *Ibid.*

[368] *Ibid.*

[369] Hangzhou.gov.cn. (7 June 2022). 杭州出台意见促进新电商高质量发展. http://www.hangzhou.gov.cn/art/2022/6/7/art_812262_59058755.html.

[370] Hsu, S. & Li, J. (2020). *Digital Payment Systems* (pp. 43–65). Columbia University Press.

[371] *Ibid.*

[372] *Ibid.*

- Scanning a QR code to transfer funds to the service provider.
- Using point-of-service (POS) machines.
- Making online purchases through the retail app directly.

Alipay, developed by Alibaba, plays a large role in Hangzhou's retail marketplace. Alipay was first launched in 2003 as an integrated payment platform on Taobao.com.[373] Alipay dominates China's cashless payment infrastructure as it is the only payment solution for Taobao and Tmall e-commerce platforms, two of China's largest online retail platforms.[374] Taobao and Tmall are also connected to activities in other marketplaces through their merchants. A continuous network of payments is being generated to strengthen Alipay's hold over e-commerce transactions.[375]

In 2004, the Alipay system was launched as a standalone third-party online payment platform. Online credit and debit payments were unpopular within Chinese society in the early 2000s.[376] Therefore, Alipay formed partnerships with leading Chinese banks and the state-owned postal service, allowing customers to deposit cash and replenish their accounts as they would at a regular bank.[377] A pivotal milestone for e-payment technology happened in May 2011, when the People's Bank of China (PBOC) granted third-party online payment licenses to Alipay and 26 other online payment companies,[378] cementing e-payment technology as a key player in the cashless society movement. As e-commerce gained in popularity, consumers soon adapted to cashless payment options, and most brick-and-mortar stores began accepting cashless payments. This happened alongside the rise of taxi-hailing online services, food deliveries, and bike sharing, which integrated mobile payment options such as Alipay and WeChat Pay into their systems.[379]

---

[373] Lu, L. (2018). Decoding Alipay: Mobile Payments, a Cashless Society and Regulatory Challenges.

[374] Rao, L. (19 June 2015). Alipay's US Chief Talks Expansion, Uber China Partnership and More. *Fortune*. https://fortune.com/2015/06/19/alipay-china-uber-alibaba/.

[375] *Ibid.*

[376] Wu & Gereffi (2018). *Op. cit.*

[377] *Ibid.*

[378] Lu (2018). *Op. cit.*

[379] Wu & Gereffi (2018). *Op. cit.*

Alipay's collaboration with financial services (Yu'E Bao) was launched in 2013. This expanded their range of fintech services as consumers could deposit money as investments with higher returns than China's bank interest rates, with the benefit of withdrawing at any time.[380] Yu'E Bao is run by the Tianhong Fund Corporation, which supports payments for public services, credit card repayment, phone lines, lotteries, split payments, and payments at physical stores. By 2017, Yu'E Bao had accumulated US$165 billion of assets under management to become the world's largest money market fund.[381] This creates a lock-in effect for end users who choose to use the app for its diverse range of services and convenience.[382] By 2018, there were 450 million registered users for Alipay in China, handling more than 50% of online payments. Alipay has also begun competing in the global mobile payment market alongside PayPal and Apple Pay.[383]

Alipay's platform provides a comprehensive and seamless financial solution to transform urban infrastructure into smart city solutions. This is through the platform's ability to handle millions of transactions, communications, and internet functions.[384] Resultingly, Alipay collects large amounts of their customer's data, repurposing it through Ali Cloud services. The cloud service also ensures the security of online financial transactions.[385] Customer spending habits through Alipay are tracked, providing customers with good credit ratings and the ability to rent flats and cars without a deposit amount.[386] This model of platform businesses for corporate convergence has since been adopted by other major web-based companies in China, such as Baidu and Tencent, or international companies,

---

[380] Lu (2018). *Op. cit.*

[381] Wang, J. & Doan, M. A. (2019). The ant empire: Fintech media and corporate convergence within and beyond Alibaba. *The Political Economy of Communication*, *6*(2), Article 2. http://polecom.org/index.php/polecom/article/view/97.

[382] Lu (2018). *Op. cit.*

[383] *Ibid.*

[384] Guo, J. & Bouwman, H. (2016). An ecosystem view on third party mobile payment providers: A case study of Alipay wallet. *Info*, *18*(5), 56–78. https://doi.org/10.1108/info-01-2016-0003.

[385] *Ibid.*

[386] *Ibid.*

such as Google or Amazon. This is due to the influence that platform businesses have towards consumers and user roles for economic and political purposes in China.[387]

Alipay has significantly impacted traditional banking and payment systems. Previously, China UnionPay was the primary payment system in the country where even Visa and MasterCard found difficulty in entering the market.[388] However, by 2015, Alipay had surpassed UnionPay with 139 billion yuan in earnings from its e-payment services. It also poses a threat to retail banking due to the number of customers choosing to move their money from bank accounts into e-wallets due to their competitive interest rates.[389] The services provided by Alipay have been met with some concern. This is due to the lack of coverage of depositors under state financial law. According to China's Deposit Insurance Regulation,[390] the maximum coverage of deposit insurance provided by the state is 500,000 yuan. This sum is not officially protected if citizens choose to move their money into fintech e-wallets. Financial regulators are bringing up the urgent need to educate financial consumers on these potential risks as most depositors are not aware of this clause, and they perceive Yu'E Bao as a safe substitute for traditional banking.[391]

Increased use of online forms of identification or payment methods has given rise to electronic Know Your Customer (e-KYC) systems for additional customer verification to protect citizens. This is to prevent money laundering and counter the financing of terrorism.[392] China has analogue forms of identifying a person through an issued national ID or bank account. This aligns with the efforts of financial institutions

---

[387] Wang & Doan (2019). *Op. cit.*

[388] Lu (2018). *Op. cit.*

[389] *Ibid.*

[390] The State Council, The People's Republic of China. (2015). *State Council Gazette Issue No.10 Serial No. 1513.* http://english.www.gov.cn/archive/publications/2015/04/10/content_281475086427148.htm.

[391] Lu (2018). *Op. cit.*

[392] Perlman, L. & Gurung, N. (2019). Focus note: The use of eKYC for Customer Identity and Verification and AML (SSRN Scholarly Paper ID 3370665). Social Science Research Network. https://doi.org/10.2139/ssrn.3370665.

worldwide to govern financial integrity and inclusion using KYC.[393] By combining their centralised identification with electronic accounts, they are successfully implementing the use of eKYC through their digital platforms. In 2010, China's Ministry of Industry and Information Technology (MIIT) implemented a requirement for real-name verification online.[394] While initially enforced by China's national telecoms, it was later extended to the registration of social media accounts. In 2016, all users of online payment platforms were also required to link their accounts to their identification card or bank account issued by mainland China. Alibaba's delivery system requires real name verification to mail items or sign for a package to protect users from cybercrime.[395] China's real name verification gives the state power to track content users and providers online.[396] This reduces online fraud and paves the way for increased financial inclusion by using platforms which require real name verification as electronic identification. This promotes a more efficient way to onboard customers through real-time or data-based verification of customer identification.[397]

4.5 Conclusion

The FTSC and other pilot initiatives have led Hangzhou to become an important economic and technological node, serving as a positive example for urban transformation across China.[398] Along with the shift to e-commerce, Hangzhou also exhibits a high internet penetration rate; 81.2% of its citizens have online connections.[399] The accessibility and convenience of e-commerce have created a shift in consumer shopping behaviours, leading to an expansion in online offerings, with more

[393] Gleb, A. (2015). CGD Policy Paper 074 February 2016 (Policy Paper No. 074). Center for Global Development.
[394] Wu & Gereffi (2018). *Op. cit.*
[395] *Ibid.*
[396] *Ibid.*
[397] Perlman & Gurung (2019). *Op. cit.*
[398] City Government of Hangzhou. (2011). Master Plan of Hangzhou City for the Development of the Ten Major industries (2011–2015). [Policy Report].
[399] Shanghai ATO & USDA (2018). *Op. cit.*

demand for a variety of goods.[400] Continuous policy attention and research are still needed to accelerate industry growth in an efficient and ethical manner. Another area is in retraining and upskilling the local retail industry, not just workers to support the technical and technological developments of the city's new retail economy but also industry captains need to know that smart transformation can happen very quickly.[401]

Hangzhou has achieved rapid progress over the past two years (2020–2022) after placing great emphasis on combining both traditional and modern innovation and industrial approaches. This combines a spatial-technological approach, automating aspects of the traditional retail industry while bringing physicality to online commerce.[402] Smart growth is a continuous process and challenges persist. While Hangzhou has revitalised economic and industrial practices, the city continues to explore different approaches to scale up and to address existing challenges with long-term innovation sustainability.[403] Despite the government providing continuous support and guidance towards innovations, Hangzhou's finances have been affected by the COVID-19 pandemic and there is now less emphasis on innovation and more focus on economic stabilisation and business feasibility through, e.g., growth of cold chain logistics for fresh produce, selectively funding, and supporting only core innovations for business stability and expansion.[404]

The retail environment is dynamic, not just from changing needs and demands of the consumer but also from a regulatory perspective. China has introduced many anti-monopoly regulations and legislations in 2021, targeted at tech sector industries, particularly the platform economy, which includes online commerce, Alibaba, and their subsidiaries. There are incoming policies, which address data collection practices,[405] online content, and the use of algorithms to target users. A main concern is that

---

[400] *Ibid.*

[401] 汪秘书长 (13 July 2022). *Op. cit.*

[402] *Ibid.*

[403] *Ibid.*

[404] *Ibid.*

[405] Kharpal, A. (20 August 2021). China Passes Major Data Protection Law as Regulatory Scrutiny on Tech Sector Intensifies. *CNBC.* https://www.cnbc.com/2021/08/20/china-passes-key-data-protection-law-as-regulatory-scrutiny-increases.html.

the proliferation of easy access to e-commerce would entice people into borrowing more than they can afford to pay back for their purchases.[406] It is important for smart retail development to be mindful of consumer needs and uphold ethical practices, including data collection, analysis, and mining practices.

[406] Law, E. (1 May 2021). China's Crackdown on Alibaba and Other Big Tech Firms Will Change Industry in Massive Way. *The Straits Times*. https://www.straitstimes.com/asia/east-asia/chinas-crackdown-on-alibaba-and-other-big-tech-firms-will-change-industry-in-massive.

CHAPTER 5

BOSTON: DIGITAL PLACEMAKING

5.1 City Overview

Location	Capital city of Massachusetts State, USA
Land Size	48.34 miles^2 (125.20 km^2) (2020)
Climate	Humid subtropical, with cold, snowy winters and warm summers
Geography (topology/ landscape)	Located in Massachusetts' Shawmut Peninsula with the Charles and Mystic rivers flowing across the city and then into Boston Harbour
Environmental Risks[407]	Climate change impacts, e.g., sea level rise, coastal storms, flooding, and heatwaves
Population	675,647 (2020)
GDP[408]	US$155 billion (2021)
GDP Per Capita[409]	US$90,626 (2021)
Key Urban Issues of Smart City	To transform the city's public realm into vibrant, welcoming, and inclusive spaces

[407] City of Boston. (2017). *Imagine Boston 2030.* https://www.boston.gov/civic-engagement/imagine-boston-2030.

[408] Boston Planning & Development Agency (BPDA), Research Division. (2023). *Boston's Economy 2023.* http://www.bostonplans.org/getattachment/c4bc18e1-d0a6-45dc-b26d-902078e6a491.

[409] Statista. (2022). *Per capita real gross domestic product of the United States in 2021, by metropolitan area (in chained 2012 U.S. dollars).* https://www.statista.com/statistics/248112/per-capita-us-real-gross-domestic-product-gdp-by-metro-area/.

Boston is the capital city of the State of Massachusetts in the United States, located in Massachusetts' Shawmut Peninsula with Charles and Mystic Rivers flowing across the city and then into Boston Harbour.[410] The City of Boston had a land area of 48.34 miles^2 (125.20 km^2) in 2020,[411] consisting of 24 neighbourhoods.[412] Boston's climate is characterised by cold, snowy winters and warm summers.[413] The hottest month is July, with an average temperature of 23.4°C, while the coldest is in January with a mean temperature of –1°C.[414] Between 2015 and 2020, Boston's average temperature was 11.9°C, almost 2°C warmer than the city's historical average temperature.[415] In parallel with this warming trend, Boston's winters experienced several record-setting snowfalls in recent years,[416] including the blizzard on 30 January 2022 that recorded the biggest single-day snowfall (23.6 inches) since 2003.[417]

Boston is the most densely populated city in Massachusetts, with a population density of 13,976.7 persons per square mile (5,396.4 persons/km^2) in 2020.[418] The City of Boston's population was 675,647 in 2020, increasing 9.4% from 2010 (617,594).[419] 28.1% of Bostonians were born overseas, which is twice higher than the national statistics (13.6%).[420]

[410] Onge, T.S. (2015). *Putting Boston on the Map: Land Reclamation and the Growth of a City*. Library of Congress Blogs. https://blogs.loc.gov/maps/2015/12/putting-boston-on-the-map/.

[411] United States Census Bureau. (n.d.). *Quick Facts: Boston City, Massachusetts*. https://www.census.gov/quickfacts/fact/map/bostoncitymassachusetts/LND110220.

[412] City of Boston. (n.d.). *Neighbourhoods*. https://www.boston.gov/neighborhoods.

[413] NOAA National Centres for Environmental Information. (2022). *State Climate Summaries 2022: Massachusetts*. https://statesummaries.ncics.org/chapter/ma/.

[414] NOAA National Centres for Environmental Information. (2023). *Climate at a Glance: City Time Series*. https://www.ncei.noaa.gov/access/monitoring/climate-at-a-glance/city/time-series/USW00014739/tavg/1/7/1990-2023.

[415] NOAA National Centres for Environmental Information. (2022). *Op. cit.*

[416] *Ibid.*

[417] Rousseau, M. (30 January 2022). Boston ties its single-day snowfall record with Blizzard of '22. *Boston.com*. https://www.boston.com/news/snow/2022/01/30/boston-ties-its-single-day-snowfall-record-with-blizzard-of-22/.

[418] United States Census Bureau. (n.d.). *Op. cit.*

[419] *Ibid.*

[420] *Ibid.*; United States Census Bureau. (n.d.). *Quick Facts: United States*. https://www.census.gov/quickfacts/fact/table/US/PST045222.

While Boston's population is significantly wealthier than the national average (a median household income of US$81,744 and US$69,021, respectively), the city has a higher proportion of its population in poverty than the national level (17.6% and 11.6%, respectively).[421] This indicates the prevalence of income inequality in Boston.

Boston is the economic powerhouse of the Massachusetts State and the top US city for fostering entrepreneurial growth and innovation.[422] Boston has sustained a growing GDP over the past two decades, except for the year 2020 when the city suffered economic recession from the COVID-19 pandemic.[423] Post-COVID, Boston has seen a strong economic recovery. In 2021, the city had a GDP of US$155 billion, increasing 7% from 2020 and surpassing the pre-COVID-2019 level.[424] In 2022, both the employment rate and consumer spending within the city had returned to the February 2020 level.[425] The rapid economic recovery came from those industries hardest hit by the pandemic (e.g., accommodation and food services) and core sectors that were less negatively impacted (e.g., professional and technical services, information, and healthcare and social assistance).[426] Inflation in Boston was high in 2022 (8.1%, similar to the national level, 8.6%), with price increases particularly felt in the food (10.5%) and energy sectors (22.8%).[427]

Boston is governed by the Mayor and the City Council.[428] The current mayor, Michelle Wu, was elected in 2021 and is the first Asian-American woman to serve as the Mayor of Boston.[429] The City Council is Boston's legislative body, consisting of 13 councillors: four at-large councillors

---

[421] United States Census Bureau. (n.d.). *Op. cit.*

[422] BPDA. (2021). *Boston's Economy 2021.* http://www.bostonplans.org/gctattachment/34b4a72f-68dc-447d-ba7d-06ce9b9db6c6.

[423] BPDA. (2023). *Op. cit.*

[424] *Ibid.*

[425] *Ibid.*

[426] *Ibid.*

[427] *Ibid.*

[428] Massachusetts Municipal Association (MMA). (n. d.) *Forms of Local Government in Massachusetts.* https://www.mma.org/resource/forms-of-local-government-in-massachusetts/.

[429] City of Boston. (n.d.). *Mayor Michelle Wu.* https://www.boston.gov/departments/mayors-office/michelle-wu.

that represent the entire city and nine district councillors that represent specific areas of the city. Councillors establish and enact public policy to meet the changing needs of Boston. They also review and approve the city's operating and capital budget every year, and track civic services to ensure their efficiency and cost-effectiveness.[430] Within the Boston City Council, there are 89 departments responsible for specific areas.[431] Four departments are closely involved in Boston's smart city initiatives: the Department of Innovation and Technology, the Mayor's Office of New Urban Mechanics, the Planning and Development Agency, and the Department of Transportation.

5.2 Smart City Development: Digital Equity for Social Inclusion

Boston's vision, established in Imagine Boston 2030 Plan, is to become "a place where Bostonians live in vibrant neighbourhoods, where all residents are able to participate in the city's dynamic economy, and where one of our most significant resources–our waterfront– is prepared to thrive for generations to come."[432] Boston's smart city journey can be traced to 2010 when Boston was one of the first US cities to experiment with smart initiatives.[433] Led by then Boston Mayor Thomas M. Menino, Boston's smart city vision was to become a host for innovation, instead of having innovation just generated from research firms or universities. An innovation district was developed in Boston's Seaport, offering a place for academia, companies, start-ups, business incubators, and accelerators to experiment and innovate.[434] Government-led innovation programmes were implemented in Boston through smart apps, such as BOS:311 and Where's My School Bus. These smart apps aim to improve public service delivery and encourage citizen involvement in the planning, development,

[430] City of Boston. (n.d.). *City Council*. https://www.boston.gov/departments/city-council.
[431] City of Boston. (n.d.). *Departments*. https://www.boston.gov/departments.
[432] City of Boston. (2017). *Op. cit.*
[433] Holmes, N. (2017). Is Boston the Smartest City? *Convene*. https://convene.com/catalyst/office/is-boston-the-smartest-city/.
[434] *Ibid.*

and governance of their community using technology. Sensors were also installed across the city for smart traffic management.[435]

In 2017, the City of Boston established smart city initiatives in two city-wide strategic plans, namely, Imagine Boston 2030 and Go Boston 2030. This policy approach differs from many US cities that developed a master plan specific for smart city development, such as the City of Chicago Technology Plan[436] and SmartLA 2028 for the City of Los Angeles.[437] More specifically, the Imagine Boston 2030 Plan outlines that Boston aims to become a smarter city through initiatives in three core areas[438]:

- Lay a foundation for digital equity.
- Invest in infrastructure in every neighbourhood.
- Protect Bostonians' digital privacy and security.

Seven key strategies are developed to achieve Boston's smart city vision[439]:

- Lay foundational "smart city" infrastructure.
- Take action to improve digital equity.
- Unlock new forms of public collaboration.
- Built the city for continuous learning.
- Make city services responsive to real-time data.
- Leverage CityScore,[440] a real-time online dashboard, to track the impact of city operations, policies, and initiatives based on various performance metrics.
- Make every resident's experience of the city personalised and intuitive.

---

[435] *Ibid.*

[436] Office of the Mayor. (2013). *The City of Chicago Technology Plan*. City of Chicago.

[437] City of Los Angeles. (2020). *SmartLA 2028: Technology for a Better Los Angeles*. https://ita.lacity.gov/sites/g/files/wph1626/files/2021-05/SmartLA2028%20-%20Smart%20City%20Strategy.pdf.

[438] City of Boston. (2017). *Op. cit.*

[439] *Ibid.*

[440] City of Boston. (n.d.). *CityScore*. https://www.boston.gov/innovation-and-technology/cityscore.

Focusing on the urban mobility sector, the Go Boston 2030 Plan establishes the strategy to pilot five smart mobility demonstration projects every year.[441] Smart mobility initiatives range from smart operations of transportation systems (e.g., transit line scheduling technology and smart parking metres) to the rollout of emerging mobility technologies (e.g., autonomous vehicles, electric vehicles and charging stations, ride-hailing, and bike-sharing). The plan also highlights the development of Neighbourhood Mobility microHUBs, now renamed as GoHubs!, throughout the city. GoHubs! seeks to integrate a wide range of urban mobility options, such as public transport, bike-sharing, ride-hailing, and electric vehicle charging stations, to help Bostonians make smarter urban transportation decisions.[442]

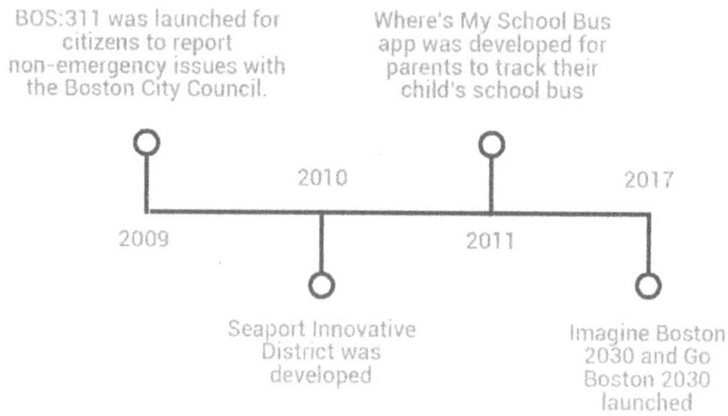

Boston Smart City Timeline

5.3 Digital Placemaking

Digital placemaking is an innovative way to support, enhance, or accelerate traditional placemaking practice through leveraging digital technologies,

[441] Boston Transportation Department. (2017). *Go Boston 2030: Imagining Our Transportation Future*. https://www.boston.gov/departments/transportation/go-boston-2030.

[442] *Ibid.*

media, and tools.[443] Placemaking takes a people-centric approach to collectively shape public spaces — especially vacant and underutilised spaces — into vibrant social, cultural, and civic places that people would like to live, work, and play in. At the heart of placemaking is the intent to create better places for people, encouraging social interaction and enhancing people's experience of the physical, economic, and social identities of a place.[444] Placemaking practices often start with small-scale elements of a place and make tactical improvements to enhance people's experience and their use of the place.[445]

Digital placemaking takes many forms. Its strategies, actions, and projects can be artistic and creative, such as interactive public artwork and events that use digital technology to create joy and wonder in familiar surroundings.[446] Digital placemaking can also be a practical solution to everyday problems like enhancing accessibility and wayfinding or helping people reimagine forgotten spaces to create further tangible connections between people and their city.

The City of Boston has been using digital placemaking strategies to advance social inclusion and transform the city's mobility systems, such as public transportation nodes, streets, and crosswalks, into inviting, vibrant, social places.[447] Digital placemaking strategies for the city's mobility system were established in two major programmes: (1) A Guide to Placemaking for Mobility and (2) the Plan for Accessible Transit Infrastructure (PATI) programme.

---

[443] Halegoua, G. & Polson, E. (2021). Exploring 'digital placemaking'. *Convergence*, *27*(3), 573–578.; Stokes, B., Bar, F., Baumann, K., Caldwell, B., & Schrock, A. (2021). Urban furniture in digital placemaking: Adapting a storytelling payphone across Los Angeles. *Convergence*, 13548565211999181.; Pang, C., Neustaedter, C., Moffatt, K., Hennessy, K., & Pan, R. (2020). The role of a location-based city exploration game in digital placemaking. *Behaviour & Information Technology*, *39*(6), 624–647.

[444] A Better City. (2016). *A Guide to Placemaking for Mobility*. https://www.abettercity.org/docs-new/Guide_to_Placemaking_for_Mobility.pdf; Loh, C. G. (2019). Placemaking and implementation: Revisiting the performance principle. *Land Use Policy*, *81*, 68–75.

[445] A Better City. (2016). *Op. cit.*

[446] Halegoua & Polson (2021). *Op. cit.*; Stokes *et al.* (2021). *Op. cit.*; Pang *et al.* (2020). *Op. cit.*

[447] Boston Transportation Department. (2017). *Op. cit.*

5.3.1 *A Guide to Placemaking for Mobility*

In 2016, the Boston Transportation Department launched the Guide to Placemaking for Mobility in partnership with A Better City, a group of business leaders from diverse industries in Boston.[448] The guide adopts a people-centric approach to transforming the public realm within Boston's mobility system and becoming vibrant, welcoming, and safe places that people would like to spend time in, instead of just spaces that people simply move through.[449] In Boston's mobility system, the public realm consists of public or civic buildings, spaces, and facilities that provide access to the mobility system, as well as various doings and beings when people travel through a city (e.g., events, performances, interactions, sounds, smells, and emotions on streets and at transit stations).[450]

The guide provides a policy framework on traditional and digital placemaking strategies and community-driven activities to reclaim streets for people. The goal is to empower planners, designers, and communities to transform sidewalks, plazas, and incidental spaces into vibrant social, cultural, and civic places.[451] The guide suggests a series of small-scale, incremental interventions in the traditional placing practice. For instance, changing people's physical movement in the mobility system through permanent adjustments to the built environment, such as reclaiming roadways for pedestrian space, sidewalk expansion, well-designed shelters and seating installed at transit stations, and surface re-designed infrastructure. Temporal interventions are also recommended, such as community events, performances, pop-up parklets, art trucks, and farmer's markets.

In addition, the guide develops digital placemaking strategies as interactive interventions to encourage social interaction and play in Boston's mobility system. This is to be achieved through harnessing digital technologies, media, and tools in the built environment, such as collaborative digital art, real-time information kiosks, and QR signages. These digital components will supplement traditional placemaking approaches to enhance

---

[448] A Better City. (2016). *Op. cit.*
[449] *Ibid.*
[450] *Ibid.*
[451] *Ibid.*

people's experience and interaction with the public realm when travelling through Boston's mobility system.[452]

5.3.2 *Plan for Accessible Transit Infrastructure (PATI) Programme*

Since 2006, the Massachusetts Bay Transportation Agency (MBTA) has been developing and implementing the PATI programme, a long-term strategic barrier-removal plan, to prioritise accessibility improvements at all bus stops and rail stations throughout the MBTA service network.[453] The PATI seeks to achieve three key objectives[454]:

- To identify all meaningful barriers to access within public-facing assets at all bus stops and rail stations,
- To establish prioritisation criteria based on community feedback,
- To develop a long-term plan for improving accessibility throughout the MBTA service network and providing capital funding recommendations.

The PATI is part of the MBTA's decades-long effort towards a fully accessible transit system. The effort dates back to 1990 when the Americans with Disabilities Act (ADA) was signed into law. The ADA Title II guarantees individuals with disabilities equal access to public transportation and establishes minimum guidelines and requirements for accessible design to ensure that buildings, facilities, rail passenger cars, and vehicles are accessible, in terms of architecture and design, transportation, and communication.[455] In response to this requirement, the MBTA

---

[452] *Ibid.*

[453] Massachusetts Department of Transportation (MassDOT). (2016). *Plan for Accessible Transit Infrastructure: Technical Support*. https://www.ctps.org/data/calendar/htmls/2016/MPO_0303_PATI_Work_Program.html.

[454] MBTA. (2017). *Plan for Accessible Transit Infrastructure (PATI). Slides presented at the Massachusetts Community Transportation Coordination Conference 2017*. https://www.mass.gov/doc/pati/download?_ga=2.121555400.1168896489.1649317963-1948041084.1649317963.

[455] ADA.gov. (2016). *Americans with Disabilities Act Title II Regulations*. https://www.ada.gov/law-and-regs/title-ii-2010-regulations/.

purchased a fleet of wheelchair-accessible buses and created an ADA-compliant Key Station Plan to guide the agency to making popular subway and commuter rail routes accessible.[456]

Accessibility improvement initiatives were accelerated by the 2006 Daniels-Finegold Settlement Agreement — a landmark agreement on MBTA's commitment to greater accessibility.[457] The agreement was precipitated by preceding events. In the early-2000s, riders with disabilities were still unable to safely or reliably access key MBTA services. Many elevators at popular stations were often out of service and bus operators were not always boarding riders with disabilities, either by choice or due to accessibility equipment malfunction. In 2002, a landmark lawsuit was filed against the MBTA by a group of 11 riders and the Boston Centre for Independent Living (BCIL). On 10 April 2006, the parties signed the Daniels-Finegold Settlement Agreement, "forever changing the future of accessibility at the T."[458] Achieving a fully accessible transit system for all was established as the MBTA's long-term vision, ensuring that riders with disabilities have "the right and ability to use public transportation in an equal, effective, and dignified manner."[459]

In 2016, the MBTA started a system-wide accessibility survey of nearly 8,000 bus stops and 250 rail stations to support the development of PATI. An innovative data-driven approach was adopted to collect, store, query, score, and report accessibility conditions of bus stops and rail stations. A tablet-based mobile application was developed to assist with the data collection. By leveraging digital data collection and management tools, this unprecedented survey work was completed within just four months without service interruption. The dataset was further developed into an open-access data platform to track the progress of accessibility improvements at bus stops and rail stations throughout the MBTA network.[460]

---

[456] MBTA. (n.d.). *History of Accessibility and the Impact of Daniels-Finegold Settlement Agreement*. https://www.mbta.com/accessibility/history.

[457] *Ibid.*

[458] *Ibid.*

[459] *Ibid.*

[460] STV. (n.d.). *Plan for Accessible Transit Infrastructure (PATI)*. https://stvinc.com/project/plan-accessible-transit-infrastructure-pati.

More recently, several digital placemaking elements were integrated into the GoHubs! pilot project in 2022. As neighbourhood mobility hubs, GoHubs! combine transportation options, information, and placemaking elements to make a multi-modal journey more convenient and enjoyable. Digital placemaking components include interactive digital information kiosks, free Wi-Fi, and digital boards for displaying local news and activities.[461] These components seek to fulfil three social objectives: (1) to enhance people's waiting and transfer experiences at Boston's major public transportation nodes, (2) to strengthen people's connection with the local community, amenities, and services, and (3) to encourage social interaction.[462]

5.4 Implementation and Impact of Boston's Digital Placemaking

Since 2010, Boston has implemented several digital placemaking initiatives across the municipality to transform public open spaces into welcoming, vibrant places for people to stay and support social interactions. This section elaborates on the implementation and impact of two of these innovations: the Mayor's Office of New Urban Mechanics and the MBTA digital signage programme.

5.4.1 *The Mayor's Office of New Urban Mechanics*

As the civic research and design team, the Mayor's Office of New Urban Mechanics (MONUM) has been a driving force in innovation and digital placemaking in Boston for over a decade. The office was established in 2010 under Mayor Thomas Menino to enable greater civic engagement and participation when the government seemed overly bureaucratic and distant from the people.[463] Since then, the MONUM has been involved in

[461] City of Boston. (n.d.). *GoHubs! Pilot Programme*. https://www.boston.gov/departments/transportation/gohubs.

[462] Boston Transportation Department. (2017). *Op. cit.*

[463] Center for Public Impact, A BCG Foundation. (2019). *Boston's Mayor's Office of New Urban Mechanics*. https://www.centreforpublicimpact.org/case-study/bostons-mayors-office-new-urban-mechanics.

a diverse array of projects across housing and education, placemaking, digital trust building, financial planning, city maintenance, road safety, food security, and more.[464]

MONUM takes a human-centric approach to smart city development, emphasising social inclusion. This approach challenges the traditional ideas about smart cities that are technocentric and heavily focused on efficiency. Since the very beginning of the office, MONUM has been exploring how innovation could make a city more enjoyable to be in, more fun and engaging to be in, and more supportive of all residents. The office centres around the human aspect of living in the city, an important issue that gets lost sometimes in the conversations about smart city efficiency.[465] In the past two to three years, MONUM has started pivoting to think about innovation as not just about fun, delight, and play but also about care. More specifically, the office explores how innovation, neighbourhoods, and infrastructure can function in a way that cares more for people.[466] MONUM published this approach in Boston's Smart City Playbook, highlighting how technology and data should be "people-centred, problem-driven, and responsible."[467]

MONUM has implemented several notable projects for digital placemaking (Figure 5.1), including the following:

- **Participatory Chinatown**: MONUM has built a character-driven virtual world that allows residents to explore Chinatown. The project was launched in 2010. Players could consider different development scenarios and think about the future shape of the neighbourhood, particularly as real master planning efforts were underway in Chinatown.[468]

---

[464] City of Boston. (n.d.). *New Urban Mechanics.* https://www.boston.gov/departments/new-urban-mechanics.

[465] Deshpande, Y. (6 July 2022). Interview with Boston Mayor's Office of New Urban Mechanics [Personal communication].

[466] *Ibid.*

[467] City of Boston. (n.d.). *Smart City Playbook.* https://www.boston.gov/departments/new-urban-mechanics/smart-city-playbook.

[468] City of Boston. (n.d.). *Participatory Chinatown.* https://www.boston.gov/departments/new-urban-mechanics/participatory-chinatown.

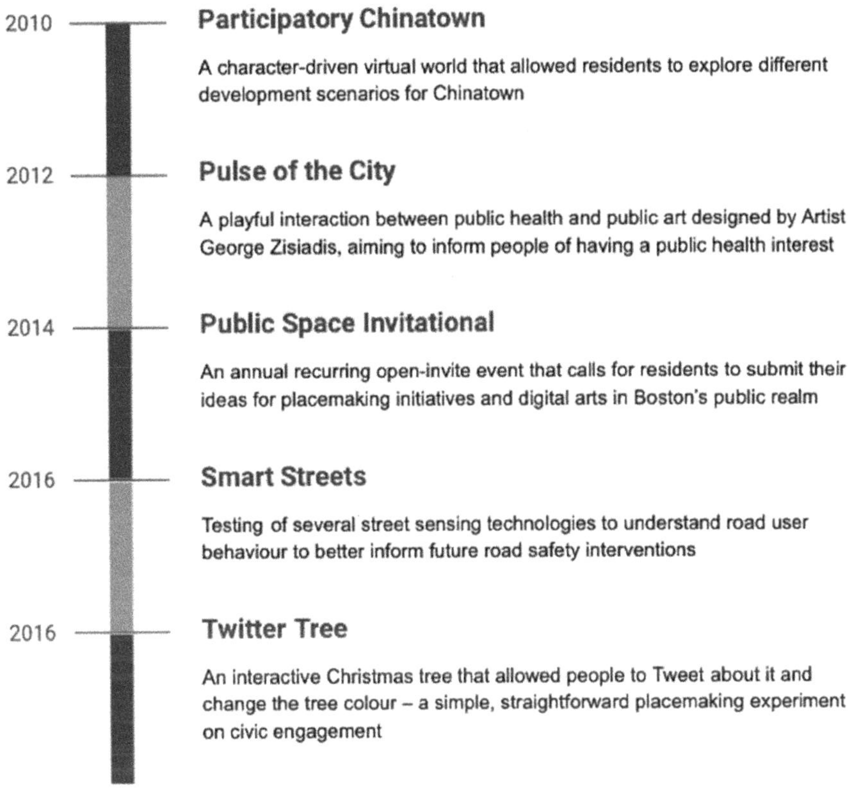

Figure 5.1. Digital placemaking projects by MONUM.

- **Pulse of the City**: In 2012, MONUM was involved in implementing the Pulse of the City project, a playful interaction between public health and public art designed by Artist George Zisiadis.[469] This project used a heart rate monitor in the shape of a heart and played musical rhymes with the user's heartbeat (Figure 5.2).[470] The art installation was implemented at five locations in the City of Boston. The project sought to inform people about having a public health interest while enjoying a playful experience with public art in public open spaces.[471]

---

[469] City of Boston. (n.d.). *Pulse of the City*. https://www.boston.gov/departments/new-urban-mechanics/pulse-city.

[470] *Ibid.*

[471] Deshpande (6 July 2022). *Op. cit.*

Figure 5.2. Pulse of the City installed in the City of Boston in 2012.
Source: George Zisiadis.[472]

This temporary prototype was installed for about nine months in Boston's public space.

- **Public Space Invitation**: Started in 2014, this annual recurring open-invite event calls for people to submit their ideas for placemaking initiatives and digital arts in the city's public realm.[473] This initiative was inspired by Philadelphia's civic design challenge. Since 2014, MONUM has considered 218 ideas with six design competitions for Boston's public spaces and has developed placemaking projects with 25 teams. Project ideas have been implemented in 12 neighbourhoods.[474]

---

[472] George Zisiadis (2012). *Pulse of the City*. http://www.georgezisiadis.com/pulse-of-the-city-2/pix7yxkbyarhbgpo7jnfxt24yo4ra3-mszbx.
[473] Deshpande (6 July 2022). *Op. cit.*
[474] City of Boston. (n.d.). *Public Space Invitational*. https://www.boston.gov/departments/new-urban-mechanics/public-space-invitational.

- **Smart Streets**: To better support the road safety initiative Vision Zero, in 2016, MONUM worked with Verizon (an American wireless network operator) to test several street sensing technologies to understand road user behaviour to better inform future safety interventions. Cameras and sensors were installed at the Massachusetts Avenue and Beacon Street intersection to understand how people use the roads and interact with the road spaces. Video and sensor data were then integrated into a data analytics platform to monitor user behaviour and traffic safety hazards. The insights helped develop interventions to enhance road safety.[475]
- **Twitter Tree**: During the 2016 Christmas holiday season, MONUM implemented an interactive Christmas tree project that allowed people to tweet about it. The central Christmas tree in City Hall was decked with 720 LED lights connected to three Wi-Fi-enabled Arduinos. People could tweet and change the tree colour using the hashtag (#WickedCoolTree).[476] This was a simple, straightforward placemaking experiment on civic engagement with positive community engagement and responses. This is a small example of placemaking initiatives that goes beyond the top-down pragmatic approach but acts as a communication channel for residents to direct their interest and love for the city.[477]

Ultimately, the ideal state of digital placemaking, as envisioned by MONUM, is to make a city that is enjoyable, fun, and engaging to visit or live in, and one that is supportive of its residents. The ideal state is to create open dialogues because residents' lives and needs are constantly changing, and so are the built and natural environments, technology, culture around the technology, and the culture of the city. Evolution is the ideal state, instead of any fixed point.[478]

---

[475] City of Boston. (n.d.) *Smart Streets*. https://www.boston.gov/innovation-and-technology/smart-streets.

[476] City of Boston. (n.d.). *Twitter Tree/Menorah*. https://www.boston.gov/innovation-and-technology/twitter-tree-menorah.

[477] Deshpande (6 July 2022). *Op. cit.*

[478] *Ibid.*

Although there are specific challenges to each digital placemaking project, some challenges are shared across different projects, including the following[479]:

- **Access to infrastructure**: Some streets in Boston do not have access to 24-hour power. As a result, several technologies that depend on access to power cannot be piloted and the challenge is to figure out alternative options.
- **Access to cross-department personnel and smoother internal workflows**: Due to its small size, MONUM often needs manpower from other departments. However, other departments may only assist depending on their availability and internal workflows. Getting the relevant expertise and commitment from different departments is a challenge. Internal engagement, communication, and partnership are crucial here to ensure that everyone is involved in the project throughout its duration.
- **Community support**: External engagement is critical to gain community support for digital placemaking initiatives and other innovations that MONUM develops and implements. The people-centric, problem-driven approach is key to ensuring that innovations can generate true value for communities.
- **Maintenance and scaling up**: Due to the nature of temporary prototypes and interventions, some digital placemaking projects lasted for only a few months. Maintenance and support were cut off after the project period. It is an ongoing challenge that MONUM seeks to solve to sustain and scale up digital placemaking innovations.

Critical success factors for MONUM's digital placemaking innovations include the following[480]:

- **Building partnerships**: Partnership is key to addressing resource access issues. Building good relationships with those who work

---

[479] *Ibid.*
[480] *Ibid.*

directly with the technology or oversee the infrastructure would help strengthen MONUM's expertise, know-how, and capacity to solve infrastructure access issues. Good relationships are also needed for working with other departments to bring innovation projects and programmes to fruition. MONUM has established collaboration with other departments and often serves as the 'connector' to bring together different departments across the city.

- **Public engagement**: Apart from the civic engagement apps and digital placemaking initiatives, MONUM holds weekly office hours for any resident in Boston to come and express their creative ideas about the city; their suggestions have inspired several innovation projects. It is also critical to cultivate relationships with community organisations, civic associations, community non-profit, and activist groups, which have on-the-ground understanding of issues in their neighbourhoods.
- **Relationships are key to maintenance too**: It helps work out long-term solutions beyond pilot projects. This requires identifying other departments, which are interested in working with MONUM in the long term and have expertise who would join and be part of the team to support and scale up the pilot project.

5.4.2 *MBTA Digital Signage Programme*

As part of the MBTA's PATI programme, the digital signage programme started in 2019 to provide riders with the most accurate real-time transit information about public transportation services.[481] The programme is led by the Customer Technology Department, the technology team within MBTA, consisting of an in-house research and development team of about 80 people.[482] The idea of digital signs and the content to display on digital

[481] MBTA. (n.d.). *Digital Signage*. https://www.mbta.com/projects/digital-signage?from=search-header-desktop&query=digital%20si.

[482] Subramanian, K. (13 July 2022). Interview with the Massachusetts Bay Transportation Authority (MBTA) (Personal communication).

screens are informed by key learning from community meetings and customer surveys, about rider needs, e.g.[483]:

- Accurate, reliable predictions of arrival times.
- Dedicated sources of real-time information that affect transit trip-planning decisions before paying the fare.
- Current trip information.

Although the digital signage programme was established in 2019, Boston's innovation pathway can be traced to 2014, which was declared "a year of action" by then US President Obama to harness technology innovations in various areas, including government services.[484] In 2017, MBTA rebuilt its website to make it more user-friendly and introduced wayfinding projects in the transit system to enhance rider journeys. The digital signage programme carried forward MBTA's efforts to make transit infrastructure and services accessible across the MBTA network through a range of accessibility technologies.[485]

Since 2019, MBTA's Customer Technology Department has developed several digital signage projects, working towards installing digital signs inside and outside rapid transit stations and at bus stops across the MBTA network. Digital signage projects implemented in the train systems include the following[486]:

- 2019 pilot of solar-powered electronic ink (E Ink) signs at Green Line stops.
- Scaling up the E Ink signs to most surface-level stops on the Green Line B, C, and E branches.
- Customer information displays (CIDs) launched in 2022 to provide dedicated real-time information about the subway system and service (e.g., alerts, delays, service disruptions, and out-of-service elevators) (Figure 5.3). CIDs will be rolled out to all subway stations in the MBTA network by 2025.

[483] MBTA. (n.d.). *Op. cit.*
[484] Subramanian (13 July 2022). *Op. cit.*
[485] *Ibid.*
[486] MBTA. (n.d.). *Op. cit.*

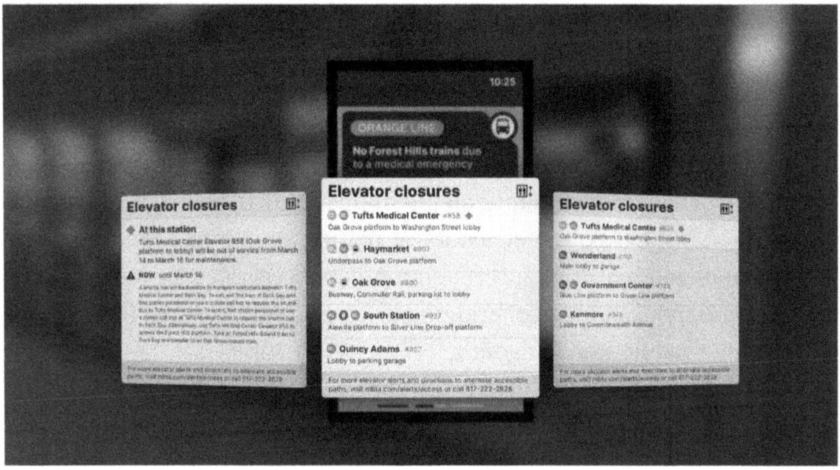

Figure 5.3. Real-time customer information display mock-up indicates out-of-service elevators in MBTA system (March 2022).

Source: MBTA.[487]

Within the MBTA bus systems, the Customer Technology Department installed new, large-format digital signs (Figure 5.4) along the centre-running bus lane on Columbus Avenue in 2021 and plans to roll out signage to future bus lane projects. The Customer Technology Department also aims to double the size of the ongoing E-Ink pilot in 2022 to provide real-time transit service information across bus stops, especially for those stops lacking access to power supply.[488]

The digital signage programme also aims to make digital signs accessible to differently abled riders. This is to be achieved by including audio-equivalent resources for riders who are sight-impaired, using large fonts to ensure readability, and factoring in viewing distance, lighting, and other environmental factors (Figure 5.5).[489]

---

[487] *Ibid.*

[488] *Ibid.*

[489] *Ibid.*

The MBTA digital signage programme faced several challenges in its innovation process[490]:

- **Mindset change**: It is not intuitive for public transit agencies to realise that they are also responsible for providing a digital layer of real-time trip data infrastructure on top of operating the hard infrastructure (e.g., trains, buses, and elevators). Creating an in-house team for rider research and real-time information provision is relevant for sensitisation and mindset change, bridging different philosophical foundations within a public transit agency.
- **Cross-functional problems**: Building in-house capacities for digital signages within a transit agency is difficult because it requires user

Figure 5.4. Digital signage installed at a new bus shelter on Columbus Avenue centre-running bus lane (November 2021).

Source: MBTA.[491]

[490] Subramanian (13 July 2022). *Op. cit.*
[491] *Ibid.*

research, procurement (which is not easy within government agencies in the US), as well as construction and maintenance of hardware and software. These elements can be different problems of their own within a government agency.

- **Evaluation**: Unlike other digital interfaces (e.g., websites), there is no digital trail regarding how riders use the real-time information on digital signages. Thus, it is not straightforward to tell whether the programme is providing people with the right information in the right place at the right time. Longitudinal survey/interview data have been collected to assess the usefulness of real-time service status in meeting rider needs.

Figure 5.5. Audio message is available from a digital signage screen at Government Center (April 2022).

Source: MBTA.[492]

[492] *Ibid.*

Critical success factors for the MBTA digital signage programme include the following[493]:

- **Problem-solving approach**: Providing riders with real-time status of public transit services to help them make more informed travel decisions.
- **User-centric**: Rider interviews and research were conducted to assist in determining the right information to be displayed at the right location within the transit system.
- **Cross-department collaboration**: Coordinate and collaborate across different departments within the MBTA to build up an accurate, real-time transit data platform.
- **In-house technology capacity**: Build a cross-functional team to ensure in-house capabilities for technology innovations.

5.5 Conclusion

Boston's digital placemaking initiatives present a people-centric approach to building a smart city towards advancing social inclusion. Digital placemaking can be implemented in various forms, ranging from public art installations that integrate interactive digital technologies (e.g., Twitter Tree and Pulse of the City) to digital signages and kiosks at public transportation nodes. Boston's digital signs are developed with all users in mind. They are equipped with audio-equivalent resources and large fonts to be more accessible to differently abled riders. A common thread underpinning Boston's digital placemaking innovations is the pursuit to bring greater values and services to the community rather than experimenting with technology solely for enhancing efficiency.

Boston's digital placemaking initiatives showcase the importance of collaboration. Building and maintaining partnerships with other departments helps overcome the issue of limited internal personnel capacity. Engagement with communities and end-users is key to ensuring that digital placemaking initiatives (and related technology innovations) are

---

[493] *Ibid.*

inclusive and respond to community needs. An important consideration is the governance model of technology innovation. Boston established a dedicated technology innovation and research team, MONUM, in 2010 when the city started its smart city journey. With a clear mission and focus on the human aspect of technology innovation, MONUM seeks to make public spaces more enjoyable and guide innovation to address real problems in the community. Against its small staff size and resource, MONUM collaborates with other developments, departments, and communities to co-design and implement digital placemaking pilots. This governance approach helps ensure that innovation responds to community needs while promoting more joined-up development in the public realm.

Research is a key task. Boston's digital signage programme is informed by in-depth user research with MBTA rides. Finding out user experiences and aspirations reinforces people-focused strategies, e.g., ensuring that the real-time information displayed on digital signs and kiosks can help riders make more informed decisions about their travels in the public transport system.

CHAPTER 6

GUANGZHOU: DIGITAL INCLUSION

6.1 City Overview

Location	Guangzhou, Capital City of Guangdong Province, China
Land Size	7434.4 km^2
Climate	Subtropical monsoon
Geography (topology/ landscape)	Mountainous terrain and alluvial plain
Environmental Risks (if any)	Water and pollution (due to industrial discharge and solid waste disposal)
Population	18,676,605 (2020)
GDP	US$346 billion (2020)
GDP Per Capita	US$21,600 (2020)
Key Urban Issues of Smart City	Ageing population, digital literacy, digital access, inclusion of rural population

Guangzhou is the capital city of Guangdong province, the third largest city in China and classified as a first-tier city.[494] Guangzhou is a key location for global transport and trade in the Guangdong-Hong Kong-Macau Greater Bay Area and the Belt and Road Initiative. It is a major science and technology innovation centre.[495] The city is one of the most densely populated areas in China, with a population density of 2,512 persons per square kilometre in 2019.[496] Its present rate of urbanisation is 87%[497] and its urban population has more than doubled since 1978. The city is situated on the Pearl River Delta in southern China, where temperatures are relatively mild in the summers and winters. Its average temperature is around 22.8°C, with an average humidity of 68%. It receives plenty of rainfall (over 1,600 mm annually), remaining evergreen through the four seasons.[498]

Guangzhou is ranked at the top of Chinese provincial-level regions in terms of economic output. The city's GDP topped 2.5 trillion yuan in 2020 (about US$388.25 billion), with an average annual growth rate of 6% since 2016.[499] Guangzhou has a longstanding economy based on international trade and is the most popular city for private investments in China.[500] Some 30,000 foreign-invested enterprises are in Guangzhou, hosting 297 of the Fortune Global 500 companies and 120 of their international or regional headquarters.[501]

---

[494] Interesse, G. (19 July 2022). Guangzhou City Profile: Industry, Economics, and Policy. *China Briefing News*. https://www.china-briefing.com/news/guangzhou-industry-economics-policy/.
Note: Cities in China are categorized into four tiers (Tiers 1–4, Tier 1 indicating cities with the largest GDP and population size. This also determines their structure of governance. Tier 1 cities are directly controlled by the central government, while the rest are divided into provinces, sub-provinces, prefectures, and counties.)

[495] *Ibid.*

[496] *Ibid.*

[497] Guangzhou Government. (2020). *Guangzhou International*. http://www.gz.gov.cn/guangzhouinternational/aboutgz/content/post_5734870.html.

[498] *Ibid.*

[499] Xinhuanet. (29 January 2021). GDP of China's Guangzhou Tops 2.5 Trillion Yuan in 2020. *Xinhuanet*. http://www.xinhuanet.com/english/2021-01/29/c_139707679.htm.

[500] UN SDG. (2021). *Op. cit.*

[501] The American Chamber of Commerce. (2018). 2018 White Paper on the Business Environment in China [Business Environment Review]. The American Chamber of Commerce.

Guangzhou is governed by a top-down administration through guiding policies issued by the Central Committee of the Chinese Communist Party (CPC) and the State Council. These policies govern regional development patterns, which are disseminated by the municipal governments of Guangzhou.[502] The People's Government of Guangzhou Municipality forms a standing committee, which oversees policy decisions and internal government operations. They implement policies by issuing administrative orders, economic plans, and financial budgets.[503] District and city mayors oversee smaller-scale civil services and neighbourhood associations.[504] Its citizens are largely concentrated in the 11 central districts of the city[505]: Yuexiu, Liwan, Haizhu, Tianhe, Baiyun, Huangpu, Huadu, Panyu, Nansha, Conghua, and Zengcheng.[506]

Guangzhou has the third largest foreign population in China[507] and the greatest number of ethnic minority residents — 56 ethnic groups with a variety of different religions and religious cultures.[508] Guangzhou faces an ageing population with an increased focus on senior care. The number of senior citizens in Guangzhou was reported to reach 1.85 million in 2020, and the government has allocated more funds towards improving senior citizen's quality of life by investing more in health and fitness, and senior social activities.[509]

---

[502] Department of Commerce of Guangdong Province. (n.d.). *Guangdong Government Policies*. http://com.gd.gov.cn/en/Policies/.

[503] Ma, X. (2021). Full text of report on the work of the Guangdong government (2021) 广东省人民政府门户网站 [Report on the Work of the Government]. 13th Guangdong Provincial People's Congress. http://www.gd.gov.cn/zwgk/zfgzbg/content/post_3185894.html; Guofa. (2018). Notice of the State Council on Issuing the Development Plan for the New Generation of Artificial Intelligence (Guofa [2017] No. 35) Government Information Disclosure Column (Technology, Education No. 35; State Development 2017). State Council. http://www.gov.cn/zhengce/content/2017-07/20/content_5211996.htm.

[504] Encyclopedia Britannica (n.d.) *Op. cit.*

[505] UN Habitat. (2021). *Vibrancy, Inclusion, and Openness; a Mega-city's Road Towards Green Development (Sustainable Development Goals)* [Local Review]. https://unhabitat.org/sites/default/files/2021/06/guangzhou_2021_en.pdf.; Encyclopedia Britannica (n.d.). *Guangzhou | History, Population, & Facts*. https://www.britannica.com/place/Guangzhou.

[506] Guangzhou Government (2020). *Op. cit.*

[507] *Ibid.*

[508] *Ibid.*

[509] The American Chamber of Commerce. (2018). 2018 White Paper on the Business Environment in China [Business Environment Review]. The American Chamber of Commerce.

6.2 Smart City Development: Sustainable Infrastructure

Guangzhou's smart city strategy focuses on incorporating new technological innovations to develop a smart, liveable, and inclusive city with a sustainable economy.[510] The city embraces the technological expansion of global trade, acting as a pilot free trade zone for cross-border e-commerce, outsourced services, and bonded logistics.[511] The city's international transport system also supports efficient international shipping and aviation-based transport.[512] The city has increased its investments in biotech, innovation and research, high-end manufacturing, new energy, and automobiles,[513] hosting 11,000 national high-tech enterprises, the third largest in China.[514] This economic expansion is supported by urban innovations that include mixed-use industrial parks, which combine housing, transport, and worksites for its employees.[515] At present, government investment drives a top-down policy approach by the Ministry of Housing and Urban–Rural Development (MOHURD), which influences the market expansion and trajectory of private sector innovations.[516]

The smart city strategy was preceded by plans for sustainable development in Guangzhou, first drafted in 1996 when departments, experts, and scholars jointly drafted the Guangzhou Agenda 21, detailing guidelines, key areas, and policy measures for Guangzhou's sustainable development. This was followed in 2000 by a strategic plan for urban development to address surrounding mountains, waterbodies, cities, cropland, and sea.[517] Guangzhou currently has several ongoing

---

[510] Ma (2021). *Op. cit.*

[511] Guangzhou Government (2020). *Op. cit.*

[512] *Ibid.*

[513] UN Habitat (2021). *Op. cit.*

[514] Guangzhou City Profile: Industry, Economics, and Policy. (19 June 2019). *Op. cit.*

[515] *Ibid.*

[516] Atha, K., Callahan, J., Chen, J., Drun, J., Green, K., Lafferty, B., McReynolds, J., Mulvenon, J., Rosen, B., & Walz, E. (2020). *China's Smart Cities Development* (p. 119). SOS International LLC, Commissioned by U.S.-China Economic and Security Review Commission.; Wan, B., Ma, R., Zhou, W., & Zhang, G. (2015). Smart city development in China: One city one policy. *ZTE Communications, 13*(4). https://doi.org/10.3969/j.

[517] UN Habitat (2021). *Op. cit.*

plans to implement smart sustainable transformation. The overarching Guangzhou Urban Plan (2017–2035) drives development under China's smart city mandate. The Double Growth Plan (2021) was proposed to further their socio-economic development, and the New Digital Infrastructure Action Plan (2020–2022) promotes the technological empowerment and digital transformation of industries in the city:

- **Guangzhou's Urban Plan (2017–2035)** aims to strengthen city organisation, planning, and regional cooperation.[518] The city has pledged to continue its sustainable efforts through technological advancement,[519] following the development of a strategic plan drafted in 2000.[520] This plan targets the development of technology and innovation clusters, linking Guangzhou with its neighbouring cities. E-platforms have also been developed to cater towards social and administrative services. For example, e-Gov Cloud, Regional Health Platform, and Digital Education City are implemented to improve the city's administrative efficiency, quality of healthcare, and education.[521]
- **Double Growth Plan (2021)**[522] proposes to combat the challenges and deficiencies that still exist due to imbalanced and inadequate social and economic development.[523] Guangzhou will fund initiatives to improve the research and development sector. To increase the effectiveness of commercial R&D, initiatives are being launched to upskill existing talents, such as engineers or specialised personnel. New schemes are also in place to standardise ethical practices and improve the social domain of innovation industries.[524] As an extension of this, universities and research institutes are encouraged through supporting policies to commercialise their research. This diversifies the evaluation of researchers beyond their publication record,

[518] Guangzhou City Profile: Industry, Economics, and Policy. (19 June 2019). *Op. cit.*
[519] Cui, Y. (2017). *Development and Practice of Smart City in China.* China Academy of Information and Communication Technology.
[520] UN Habitat (2021). *Op. cit.*
[521] Cui (2017). *Op. cit.*
[522] Ma (2021). *Op. cit.*
[523] UN Habitat (2021). *Op. cit.*
[524] Ma (2021). *Op. cit.*

Guangzhou Smart City Timeline

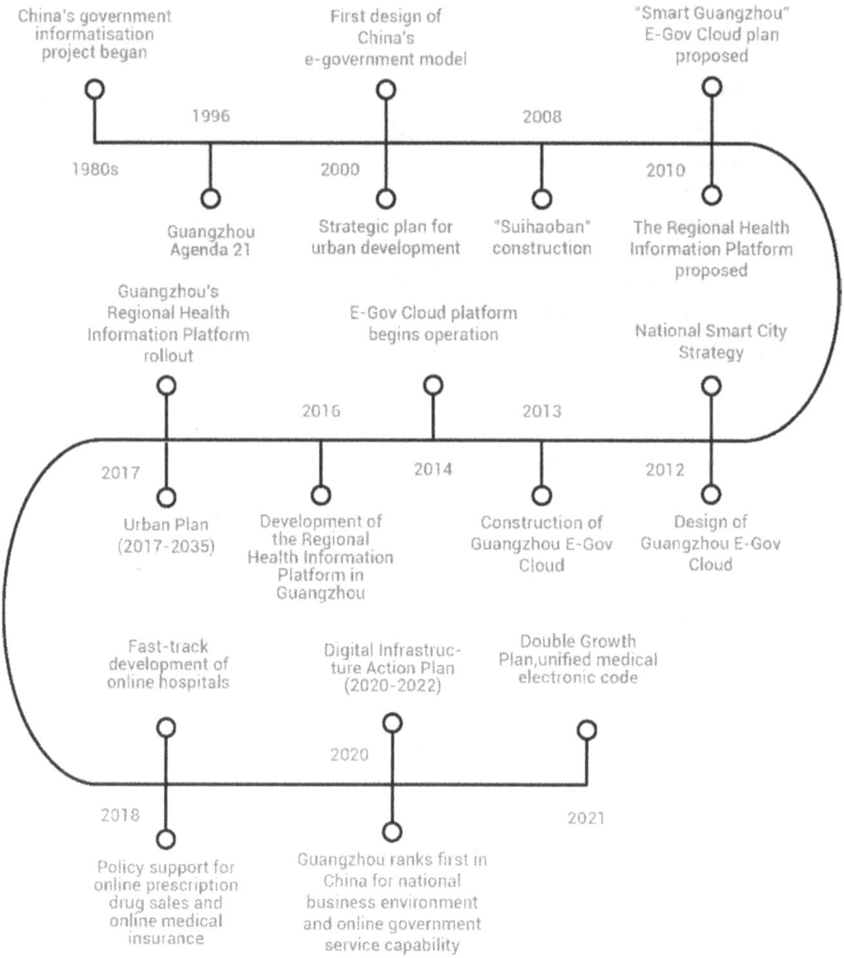

to consider commercial achievements through the number of patents or the translations of their research. The Guangzhou government has allocated a budget of 10 billion yuan in 2017 for supporting subsidies. Enterprises buying scientific and technological patents can seek government subsidies for 5% of the research value. New sci-tech companies can also apply for subsidies of 50% of their rent for the first 3 years.[525]

[525]Yu, F. & Fu, Q. (2016). Spotlight on Guangzhou. *Nature*. https://doi.org/10.1038/nj0483.

- **New Digital Infrastructure Action Plan (2020–2022)** promotes technological empowerment and digital transformation of industries in the city. This is to accelerate development leading to new digital infrastructure and stimulate regional economic vitality. This is implemented through 40 new digital infrastructure projects, focusing on four major areas: 5G network construction, cross-border integration of artificial intelligence and big data, industrial internet integration and innovation, and upgrading smart charging infrastructure for new energy automobile industry.[526] This is supported by a total investment of more than 30 billion yuan, with plans to scale the artificial intelligence industry to more than 120 billion yuan by the end of 2022.[527]

6.3 Digital Inclusion

Digital inclusion plays a large role in Guangzhou's smart city vision. The aim is to make government services more efficient and accessible to citizens through upskilling and expanded use of e-platforms.[528] Healthcare sectors and industrial economies will create more accessible and supportive industries, e.g., advancements in healthcare, education, and social administration through e-platforms to make them more accessible to citizens with improved efficiency.[529] Digital inclusion also encompasses services catering towards minority groups. At the national level, the Chinese Disabled Person's Federation highlighted the potential of the internet for people with disabilities in 2007. This led to the first targeted action to initiate digital inclusion during the implementation of the

[526] Yeung, G. & Lu, C. (2020). Comparison of Smart City Policies and Practices Among Hong Kong, Macau, Guangzhou and Shenzhen in the Greater Bay Area in South China. In *2020 5th International Conference on Universal Village (UV)*, 1–6. https://doi.org/10.1109/UV50937.2020.9426224; South China Morning Post. (11 January 2021). *What are Guangzhou's tech Mega Projects and What Role Will They Play in Bay Area?* https://www.scmp.com/business/china-business/article/3117213/how-guangzhous-tech-innovation-mega-projects-aim-bolster.

[527] Zheng, C. (8 July 2020). *Guangzhou to Speed Up Smart City Construction: Official.* https://www.chinadaily.com.cn/a/202007/08/WS5f05c9efa310834817258263.html.

[528] Guangzhou Government (2020). *Op. cit.*

[529] *Ibid.*

Eleventh Five-year Plan of China's Disability Work (2006–2010). However, it was noted that online information services for people with disabilities were basic. This was rectified in 2015 when the Internet Plus Action Plan proposed to utilise ICTs in systematic ways for people with disabilities' rehabilitation, education, employment, poverty alleviation, social support, and social security.[530]

Big ICT companies such as Alibaba, Tencent, and Baidu play a large role in aiding people with disabilities under the 2015 Internet Plus guidelines. Alibaba provides accessible retail options through Taobao e-commerce, while Baidu through their search engine provides information on disability and health-related information. The result is a material enclave and an alternative society for people with disabilities in China, acting as a heterotopia within Chinese society.[531]

Guangzhou's strong urban infrastructure is a key factor in its smart city developments. There is strong financial and industry support for the information industry, which has become a pillar industry of Guangzhou's economy.[532] The China Industry and Planning Research Institute's assessment of Guangzhou's digital information infrastructure (measuring the support provided by Guangzhou's key pillars, such as economy, commerce, government administrative systems, public services, and key technology providers) indicates that 72% of enterprises have developed e-commerce activities. E-government systems cover 80% of administration and public services; telephone and internet access rates are at 230.7% and 72.9%, respectively.[533] The city's population also has high levels of digital literacy and access to online services, quantified under "online to offline" and "offline to online" (O2O) data, measured by online order generation and offline delivered products and services. In 2015, Guangzhou ranked second in China, behind Beijing, with comprehensive O2O development. This is due to the city's strong industrial support capabilities with a high user penetration rate and high probability to convert non-O2O users

[530] *Ibid.*

[531] Lin, Z., Yang, L., & Zhang, Z. (2018). To include, or not to include, that is the question: Disability digital inclusion and exclusion in China. *New Media & Society, 20*(12), 4436–4452.

[532] Cui (2017). *Op. cit.*

[533] *Ibid.*

into future users. Resultingly, many e-commerce platforms are available in Guangzhou, supporting cross-border trade and online food services.[534]

6.4 Implementation and Impact of Digital Inclusion Initiatives

This section elaborates on the implementation and impact of two of these innovations: e-gov cloud and regional health information platform.

6.4.1 *Suihaoban and e-Gov Cloud*

Guangzhou City is supported by the all-in-one mobile app "Suihaoban" (穗好办), which consolidates all municipal services, acting as a main citizen-facing portal of mobile government affairs in all municipal departments and districts.[535] Services include social security, medical and health care, education, transportation, household registration, housing provident fund, real estate, labour and employment business, legal aid, and the establishment and operation of enterprises.[536] Its construction began in 2008, unifying government affairs services for reservations, inquiries, and handling of administration for citizens and enterprises.[537] By 2022, this

[534] HKTDC Research (March 2020). *Op. cit.*

[535] Guangzhou Government Affairs Service Data Administration. (4 August 2022). How Smart Governance Promotes Social Inclusion in Guangzhou [Personal Communication].; Guangzhou Municipal Finance Bureau. (2020, April 26). "穗好办": 广州政务服务最好办 广州市财政局网站 ("Suihaoban": Guangzhou government services are the best. Guangzhou Finance Bureau website). http://czj.gz.gov.cn/zwgk/zfxxgkml/gzdt/content/post_5807340.html.

[536] 新快报. (13 August 2022). 广州公布3334项"穗好办"政务服务事项清单 (Xin Kuai Bao. Guangzhou publishes a list of 3,334 "Easy-to-use" government services). https://finance.sina.com.cn/tech/roll/2022-08-13/doc-imizirav7935094.shtml.

[537] 广州市政务服务数据管理局. (25 May 2020). 在穗创业者开办企业"再提速", "穗好办" *APP* 为你带来新体验 — 广州市增城区人民政府门户网站 (Guangzhou Municipal Government Service Data Management Bureau. (25 May 2020). The process of starting a business in Guangzhou has been accelerated. The "Suihaoban" APP brings you a new experience — Guangzhou Zengcheng District People's Government Portal). http://www.zc.gov.cn/zfxxgkml/gzszcqzwfwsjglj/qt/content/post_5849615.html.

platform has hosted 2,800 services with intelligent search functions and helpdesk functions. To increase accessibility, 5,000 government self-service kiosks have been set up for citizens without access to the mobile app, covering all districts, towns, and villages.[538]

Suihaoban has built a network and virtual map of government and public service resources, comprising institutional service station providers and community spaces and facilities.[539] There are plans to expand to an online catalogue of medical treatments, cultural and sports events, tourism, and accommodation available in Guangzhou providing a one-stop "smart life" service scenario. Enterprise service capabilities are to be optimised using authenticated electronic signatures, seals, and other support capabilities to fully digitise administrative processes while ensuring cyber security.[540] This creates a documented digital trail of market access, investment and construction trends, credit transactions, production, operations, and market processes. Analysis of enterprise-related data will be made possible creating a deeper understanding of the economic ecosystem and increasing the city's ability to accurately support private services, create new innovative service models, and create an interconnected network between governments and businesses.[541]

At the backend of Guangzhou's municipal government administrative platform is e-Gov Cloud database, which has been in operation since 2014. This platform supports more than 1,900 business systems of over 500 state-owned or public organisations in the city.[542] This allows for the consolidation of digital administration regarding education, social insurance, taxation, medical care, enterprise registration, and business licensing across the country (Figure 6.1).[543] The Guangzhou Municipal Government (GMG) proposed its "Smart Guangzhou" e-Gov Cloud plan in 2010. The design began in 2012 and construction in 2013, adopting China's existing government cloud platform for all bureaus, commissions,

---

[538] 新快报 (Xin Kuai Bao) (13 August 2022). *Op. cit.*

[539] 广州市政务服务数据管理局 (Guangzhou Municipal Government Service Data Management Bureau) (25 May 2020). *Op. cit.*

[540] *Ibid.*

[541] 新快报 (Xin Kuai Bao) (13 August 2022). *Op. cit.*

[542] Guangzhou Government Affairs Service Data Administration (4 August 2022). *Op. cit.*

[543] Ma (2021). *Op. cit.*

and offices under them.[544] While this platform is more recent, China's Office Automatization Project dates back to the late 1980s and envisioned to computerise office administrative work, building an internal network for government officers.[545] The e-government model was designed at the end of 2000 by a team led by experts from the Ministry of Science and Technology, the State Informatisation Office, and the Ministry of Public Security.[546] This model is applied at all levels, promoting an e-government platform for internal administration, resource management, macro-economy management, taxation, finance, identification cards, auditing, public and social security, agriculture, and natural resource management.[547] This simplifies the procedures for administrative affairs by storing consolidated information while also calculating and analysing data to reduce administrative decision-making time.[548]

Guangzhou's e-Gov Cloud follows the State Council's recommended construction guidelines[549] and is supported by a platform known as the GMG Information Cloud (GMGIC), optimised by Huawei. The project consists of a cloud computing centre, a cloud industry operation centre, an innovation demonstration centre, and a cloud development platform, promoting the urban integration of Guangzhou.[550] It provides support services, such as computing, storage, network, security, and basic software.[551]

---

[544] Yu, Z. & Zhang, D. (2017). Shaping Urban Perfection in Guangzhou with Government Cloud — Huawei Publications. *Huawei*. https://www.huawei.com/en/technology-insights/publications/winwin/30/guangzhou-government-information-cloud.

[545] Zhang, J. (2006). Good governance through e-governance? Assessing China's e-government strategy. *Journal of E-Government, 2*(4), 39–71.

[546] *Ibid.*

[547] *Ibid.*

[548] Wang, J. (12 November 2018). Tech Giant Blazes Trail for Profound Digital Change. *China Daily*. https://www.chinadaily.com.cn/cndy/2018-11/12/content_37242235.htm.

[549] Han, G. (2017). Notice of the General Office of the State Council on Issuing the Guidelines for the Construction of the 'Internet + Government Service" Technical System (Guoban Han [2016] No. 108) Government Information Disclosure Column (Comprehensive Government Affairs\e-Government Affairs 000014349/2016-00274; State Office Letter (2016) No. 108). Office of the State Council. http://www.gov.cn/zhengce/content/2017-01/12/content_5159174.htm.

[550] E-Guangzhou. (13 September 2017). *Huawei launches cloud computing project in Guangzhou*. http://www.eguangzhou.gov.cn/2017-09/13/c_103763.htm.

[551] Yu & Zhang (2017). *Op. cit.*

Huawei's unified management platform has consolidated GMG's online data to reduce data overlaps, server usage, storage, and hardware by 75%. Additionally, it has also shortened the timeline of project implementation by over 70%.[552] GMG has effectively built its own government cloud service marketplace with flexible components, which can quickly integrate emerging technology and new customer requirements.[553] Guangzhou's e-Gov Cloud platform ranks 5th in China on the open data platform index, indicating high levels of informatisation in their government affairs.[554]

Since the implementation of the e-Gov cloud, there have been several main benefits of this consolidated platform. The overall civil services in Guangzhou have been improved to ensure efficiency and reliability in regulating social security and financial equality. The platform utilises the information of social security cards to combine with household financial data to conduct status checks on household economic statuses.[555] The system handles automatic data reconciliation between government and industry, which includes information, such as insurance and real estate registrations. This plays a major role in tackling illegal activities, such as false declarations and false documentation.[556] This has effectively reduced tax fraud while increasing the tax collection rates from 69% to 98% after deployment in 2017. It has also facilitated data collection to analyse the city's economic performance and operations with more accuracy and detail.[557] The government is able to check on the status of low-income families that apply for income support. Before 2016, this process was tedious due to gaps in consolidated personal information and background, resulting in false claims for welfare payments.[558] Citizens are now only required to present a single license when applying for government services.[559]

---

[552] *Ibid.*

[553] *Ibid.*

[554] Cheng, H., Zhao, B., & Liu, X. (2019). Study on development bottleneck and service strategy of China e-government Cloud platform in the big data era. *IOP Conference Series: Materials Science and Engineering*, *490*, 062068. https://doi.org/10.1088/1757-899X/490/6/062068.

[555] Cui (2017). *Op. cit.*

[556] Wang (12 November 2018). *Op. cit.*

[557] Yu & Zhang (2017). *Op. cit.*

[558] *Ibid.*

[559] *Ibid.*

As of 2021, the objectives for the GMGIC platforms have been achieved with 100% of all government departments migrated to the cloud platform.[560] The municipality is on track for further expansion into next-gen technologies such as cloud computing, big data, and artificial intelligence in its government IT framework.[561] Planned advancements include the construction of a data centre to support cloud infrastructure of GMG. Data sharing is being expanded between provincial and municipal platforms, which includes the shared application of integrated government service platforms. This action plan is given a 3-year deadline beginning in 2021 which plans to achieve the complete integration of the overall administration on one platform.[562]

The implementation of Suihaoban and e-gov cloud encountered challenges upon initiation due to the lack of a unified data share mechanism and the absence of an interconnected cross-sector government information system.[563] There are also issues with high operation and maintenance costs, arising from repeated purchases and increased R&D costs. The result is "information isolated island" and "data stack" due to repetition of data, poor consistency in formatting, and low utilisation, which negatively impacts government service levels and efficiency.[564]

Front-end platforms also lacked proper service design in the beginning, resulting in convoluted functions and missing information on the portal, with blank pages and malfunctioning apps.[565] To solve these challenges, researchers suggested to establish the online government service based on cloud computing framework collaboration to construct a cross-regional, cross-sector, and multi-level government data platform that is openly shared within the public service.[566] Huawei's cloud computing unified platform service was thus introduced and resolved these issues.[567] However, improve-

[560] Ma (2021). *Op. cit.*; Yu & Zhang (2017). *Op. cit.*

[561] *Ibid.*

[562] Ma (2021). *Op. cit.*

[563] Fang, Y. & Chen, L. (2008). The design and research of basic information platform about the electronic government in Guangzhou. *2008 International Seminar on Business and Information Management, 2*, 525–527.

[564] Cheng *et al.* (2019). *Op. cit.*

[565] *Ibid.*

[566] *Ibid.*

[567] Yu & Zhang (2017). *Op. cit.*

ments can be made for this initiative to be more open and predictive of community engagement and need to allow the co-production of services, financial investment, automation, collaboration, and governance.[568]

6.4.2 *Regional Health Information Platform*

Guangzhou's Regional Health Information Platform collects and stores electronic records from healthcare institutions for residents and medical professionals to access patient records, prescriptions, and other medical information. It began development in 2016, under the Ministry of Science and Technology as a centralised integrated data platform.[569] Its original purpose was to collect and store patient data and bio-samples to create a population database.[570] The platform was rolled out in different regions of China following the same model to streamline healthcare services. The motivation was a government health reform in 2010 to build a regional medical consortium and enhance the efficiency of health information exchange.[571] The regional consortium would be formed by provincial, municipal, county, and village medical services through a vertical hierarchy of referral, or a horizontal hierarchy of different medical specialties. This move is supported by government-level policies to enhance online services and promote market development of online hospitals.[572]

The initial development in 2016 coincided with the implementation of big data in the healthcare sector.[573] In April 2018, Internet +

---

[568] *Ibid.*

[569] Zhang, L., Wang, H., Li, Q., Zhao, M.-H., & Zhan, Q.-M. (2018). Big data and medical research in China. *BMJ*, j5910. https://doi.org/10.1136/bmj.j5910.

[570] *Ibid.*

[571] Liang, J., Zheng, X., Chen, Z., Dai, S., Xu, J., Ye, H., Zhang, Z., Ge, F., & Lei, J. (2019). The experience and challenges of healthcare-reform-driven medical consortia and Regional Health Information Technologies in China: A longitudinal study. *International Journal of Medical Informatics, 131*, 103954. https://doi.org/10.1016/j.ijmedinf.2019.103954; Li, X., Lu, J., Hu, S., Cheng, K., De Maeseneer, J., Meng, Q., Mossialos, E., Xu, D. R., Yip, W., Zhang, H., Krumholz, H. M., Jiang, L., & Hu, S. (2017). The primary health-care system in China. *The Lancet, 390*(10112), 2584–2594.

[572] Deloitte (2021). Internet Hospitals in China: The New Step into Digital Healthcare (p. 28).

[573] The State Council, PRC (2016). [Policies]. *China to Boost Big Data Application in Health and Medical Sectors*. http://english.www.gov.cn/policies/latest_releases/2016/06/24/content_281475379018156.htm.

Healthcare development was promoted and fast-tracked the development of online hospitals. Offline medical organisations were encouraged to provide online services, reinforcing the legal status and regulations of online healthcare.[574] Subsequent policy revisions were published in late 2019, which allowed for the online sale of prescription drugs[575] and implemented medical insurance coverage for online services.[576] Insurance coverage was expanded in March 2020 to promote direct payment by insurance for prescription medicine and online medical treatment for return patients.[577] Online and offline models of healthcare, and the digitisation of online treatment, electronic prescriptions, and drug distribution were also promoted in the same year.[578] Another key policy was for all public hospitals to join medical partnership systems by 2020.[579] Since then, many academic research projects have received approval to evaluate online national datasets of patient records,[580] paving the way for detailed research on the clinical and economic burden of chronic diseases.[581]

---

[574] The State Council, PRC (28 April 2018). State Council issues guideline on Internet Plus healthcare. http://english.www.gov.cn/policies/latest_releases/2018/04/28/content_281476127312948.htm.

[575] The State Council, PRC. (n.d.a). Laws & regulations — Archive. http://english.www.gov.cn/archive/lawsregulations.

[576] The State Council, PRC. (n.d.b). Archive. http://english.www.gov.cn/archive.

[577] Deloitte (2021). *Op. cit.*; The State Council, PRC (n.d.). *Op. cit.*

[578] Deloitte (2021). *Op. cit.*

[579] *Ibid.*

[580] Zhang *et al.* (2018). *Op. cit.*

[581] Zhang, L., Long, J., Jiang, W., Shi, Y., He, X., Zhou, Z., Li, Y., Yeung, R. O., Wang, J., Matsushita, K., Coresh, J., Zhao, M.-H., & Wang, H. (2016). Trends in chronic kidney disease in China. *New England Journal of Medicine, 375*(9), 905–906.; Zhou, M., Wang, H., Zhu, J., Chen, W., Wang, L., Liu, S., Li, Y., Wang, L., Liu, Y., Yin, P., Liu, J., Yu, S., Tan, F., Barber, R. M., Coates, M. M., Dicker, D., Fraser, M., González-Medina, D., Hamavid, H., Hao, Y., Hu, G., Jiang, G., Kan, H., Lopez, A. D., Phillips, M. R., She, J., Vos, T., Xu, G., Yan, L. L., Yu, C., Zhao, Y., Zheng, Y., Zou, X., Naghavi, M., Wang, Y., Murray, C. J. L., Yang, G., Liang, X. (2016). Cause-specific mortality for 240 causes in China during 1990–2013: A systematic subnational analysis for the Global Burden of Disease Study 2013. *The Lancet, 387*(10015), 251–272.

Figure 6.1. Guangdong unified electronic health code app.
Source: Guangzhou Government.[583]

Guangzhou's platform has adapted this centralised integrated platform model into a unified appointment registration system with an accompanying smart app that provides online medical services. Services include appointment-making, queues, payments, direct communication with healthcare workers, and access to personal reports (Figure 6.1).[582] As of 2017, Guangzhou's Regional Health Information Platform has been applied in five municipal hospitals and five grassroots medical institutions. It has built over 8 million electronic health records and the platform is being promoted to all municipal and district hospitals.[584] By 2023, the platform has further integrated Guangzhou's population records into its online database. This is supported by the 2021 plan to provide a unified

[582] Cui (2017). *Op. cit.*
[583] Guangzhou Government (17 March 2021). *Op. cit.*
[584] *Ibid.*

electronic code for all Guangdong citizens across 1,727 hospitals in the province, extending to other Greater Bay areas, including Guangzhou, Shenzhen, Zhongshan, Jiangmen, and Zhaoqing.[585] This would allow citizens to share their personal health records and visit different hospitals in Guangdong without manual administration. Citizen "ID cards" are also linked to this unified health code. Citizens and foreigners can either use their local ID cards, or a registered valid ID, to make appointments, run tests, get consolidated test results, and obtain prescriptions and medications while making payments online.[586]

The Regional Health Information Platform has led to the development of the internet hospital, which provides online healthcare to citizens who are disadvantaged by their location, lacking access to quality medical care. Medical consultation facilities have been established in rural areas to connect rural patients to doctors based in a top-level hospital in Guangzhou via the internet hospital platform. The patient can send images of their prior health checks or do a variety of diagnostic tests through machines on site. The doctor would then make an online diagnosis, prescribing medication, which the patient will obtain the label for to buy drugs at the consulting facility or drug store or have it sent to their doorstep.[587]

E-platforms and new cloud systems have brought various advantages to the healthcare sector. The first is increased healthcare accessibility for populations including rural residents by improving their accessibility, efficiency, and quality of healthcare through online access to major hospitals in the region while consolidating databases for medical records and research.[588] China has seen the rapid digitisation of medical services between 2019 and 2021. In May 2019, the number of internet hospitals

---

[585] Guangzhou Government. (17 March 2021). Guangdong launches unified electronic health code for 1727 hospitals. http://www.gz.gov.cn/guangzhouinternational/home/citynews/content/post_7143042.html.

[586] *Ibid.*; Tu, J., Wang, C., & Wu, S. (2018). Using technological innovation to improve health care utilization in China's hospitals: The emerging 'online' health service delivery. *Journal of Asian Public Policy, 11*(3), 316–333.

[587] Tu, J., Wang, C., & Wu, S. (2015). The internet hospital: An emerging innovation in China. *The Lancet Global Health, 3*(8), e445–e446.

[588] Han, Y., Lie, R. K., & Guo, R. (2020). The internet hospital as a telehealth model in China: Systematic search and content analysis. *Journal of Medical Internet Research, 22*(7), e17995. https://doi.org/10.2196/17995; Tu *e al.* (2018, 2021). *Op. cit.*

reached 160, while by March 2021, that number grew to 1,100. Additionally, 7,700 hospitals have set up online registration and diagnosis, while over 2,200 tertiary public hospitals (the highest grade) have progressed to information connectivity and sharing.[589] Regulatory mechanisms have been established in 30 provincial-level regions for the establishment of online medical services to ensure the safety of patients. For oversight, the National Health Commission will begin to release uniform standards for managing health codes across regions and push for integrating COVID-19-related services, such as nucleic acid tests, vaccination status, and travel histories.[590]

Second, online health platforms provide a much cheaper alternative to outpatient care than traditional hospital visits. Online prescriptions cost a quarter of the cost of drugs from a top-level hospital.[591] As a result, the internet hospital model would have significant implications for providing quality health care in developing and remote regions. This system could also improve the overall health education and raise health awareness among the wider public due to the ease of access to professional health consultancy services.[592]

One of the major challenges with implementing electronic medical records in China was that healthcare data were previously unstructured. In 2016, over 90% of hospitals in China used electronic records, which were difficult to access with poor quality data.[593] This has impeded the establishment of a unified system, especially since hospitals were previously supported by over 300 different commercial providers of electronic platforms. Each platform varied in structure and data standards, resulting in records being isolated to individual establishments.[594] China also lacked consistent implementation of medical terminology. Despite the mandated implementation of the International Classification of Diseases (ICD) by the National Health and Family Planning Commission in 2002, the use of different medical platforms has resulted in different coding of terminolo-

[589] Wang, X. (24 March 2021). Strides taken toward online medical services. *Chinadaily.com.cn*. http://www.chinadaily.com.cn/a/202103/24/WS605a90c6a31024ad0bab12d7.html.

[590] *Ibid.*

[591] Tu *et al.* (2015). *Op. cit.*

[592] *Ibid.*

[593] Zhang *et al.* (2018). *Op. cit.*

[594] *Ibid.*

gies. Widely accepted terminology systems, such as the SNOMED CT, UMLS, or GALEN, are not available in China.[595] Therefore, efforts also had to be made to promote a universal standard for biomedical information, services, and health records.[596]

Another challenge relates to poor physical access to high-quality health, which care can be attributed to patients seeking help at high-level hospitals even for minor ailments and overcrowding the hospitals. Although these ailments can be treated in primary health-care institutions at affordable prices, many patients are unwilling to go to them when given a choice due to a lack of confidence in the professional skills and quality of care provided. Skilled doctors are also often unwilling to work at the community level in rural areas for financial and professional reasons.[597] While internet hospitals provide a feasible solution to meet demands for high-quality outpatient services, considerations must be made to improve the overall quality of healthcare. It is important to improve the level of trust in local healthcare institutions, and for social inclusivity to be met, both inpatient and outpatient care regardless of location.[598]

While online platforms in Guangzhou are shown to have relatively high levels of service effectiveness, perceived ease of use, and information timeliness that meet residents' requirements of use, they also suffer from information accuracy issues, poor service responses, poor system reliability, high perceived costs, and perceived risks.[599] There were usability issues and problems with the inclusion of the regional health platform for more remote populations and older adults. Research suggests that the adoption of a mobile platform may unintentionally skew privilege towards populations with more access to online resources while exacerbating disparities for those without.[600] Urban residents have rated these apps higher for service quality and information accuracy than rural resi-

---

[595] *Ibid.*

[596] *Ibid.*

[597] Tu *et al.* (2015). *Op. cit.*

[598] *Ibid.*

[599] Lin, S., Liang, X., Zhang, B., Xing, F., & Peng, G. (2020). Investigating users attitudes and perceptions towards the usage of smart city apps. In N. Streitz & S. Konomi (Eds.), *Distributed, Ambient and Pervasive Interactions.* Springer.

[600] Magsamen-Conrad, K., Wang, F., Tetteh, D., & Lee, Y.-I. (2020). Using technology adoption theory and a lifespan approach to develop a theoretical framework for ehealth literacy: Extending UTAUT. *Health Communication, 35*(12), 1435–1446.

dents, while older adults tend to face several obstacles with the operation of smart apps.[601] Younger people tend to be more adept at adopting web-based healthcare tools and hospitals. Older people with more health issues might not be able to benefit from these new platforms as they tend to use communication technology less.[602] Health problems and healthcare access difficulties associated with age have led to concerns about the transition into smart healthcare among the older generation.[603] It is important for smart city development to regularly review and consider these user issues and put in place policies designed to assist people of all ages with the use of technology and electronic hospital access.[604] Technical guidance and support can be provided for older people through training sessions. These sessions may be conducted within hospitals or in community areas. Apps can also be made more elder-friendly by including elder-friendly features (e.g., larger font) and keeping the user interface universal and consistent. User literacy is another potential enhancement area, including keeping traditional channels for those unable to learn or utilise mobile technology.[605]

The pandemic provided a key development catalyst. The "Internet + medical care" scheme has been accelerated in Guangzhou and other regions during the COVID-19 pandemic to minimise transmission risks and to prevent over-taxation of the medical system.[606] Online healthcare

---

[601] *Ibid.*

[602] Li, P., Luo, Y., Yu, X., Wen, J., Mason, E., Li, W., & Jalali, M. S. (2020). Patients' perceptions of barriers and facilitators to the adoption of e-hospitals: Cross-sectional study in Western China. *Journal of Medical Internet Research, 22*(6), e17221. https://doi.org/10.2196/17221; Tu, J., Shen, M., Zhong, J., Yuan, G., & Chen, M. (2021). The perceptions and experiences of mobile health technology by older people in Guangzhou, China: A qualitative study. *Frontiers in Public Health, 9*, 683712. https://doi.org/10.3389/fpubh.2021.683712.

[603] Cao, M. J., Guo, X. L., Yu, H., Chen, L. Y., & McDonald, T. A. (2014). Chinese community-dwelling elders' needs: Promoting ageing in place: Elderly needs of community care in Hangzhou, China. *International Nursing Review, 61*(3), 327–335.

[604] Tu *et al.* (2021). *Op. cit.*

[605] *Ibid.*

[606] Huang, M., Wang, J., Nicholas, S., Maitland, E., & Guo, Z. (2021). Development, status Quo, and challenges to China's health informatization during COVID-19: Evaluation and recommendations. *Journal of Medical Internet Research, 23*(6), e27345. https://doi.org/10.2196/27345.

platforms have extended to provide online consultations, prescriptions, and drug deliveries, moving many hospital procedures online.[607] Web-based consultations are a new aspect of online healthcare, providing free general medical services and mental health counselling for the public through the WeChat app.[608] Specialised physicians in a wide range of healthcare services, such as internal medicine, surgery, traditional Chinese medicine, rehabilitation, and psychology, would be referred to patients and medical personnel through this online service.[609] This is linked to the GD2H Internet Hospital specialised service, which also serves to connect healthcare workers facing issues in more remote locations or in-care centres.

Professional training on COVID-19 diagnosis, monitoring, and treatment was also carried out through this platform to quickly disseminate new information. Patients could upload materials about their illness, past medical history, symptoms, and test results in different formats. These data would then be encrypted and transmitted, and patients would receive e-prescriptions, which were connected to third-party drug distribution agencies providing home delivery.[610] By the end of 2020, these automated COVID-19 services had been used by over 7 million people. The number of general web-based consultations increased from 30 visits pre-epidemic to 122 visits during the height of COVID-19, with an average of 73 visits post-epidemic. The uptake demonstrates the acceptability and feasibility of complementing offline hospital services with online outpatient care.[611]

6.5 Conclusion

Guangzhou's strategy towards the development of a smart city has been heavily focused on a well-rounded technological approach targeting improvements to social welfare and inclusion. Many initiatives are streamlining civil administration while improving the accessibility and convenience of government services, healthcare, and education. The

[607] Lian *et al.* (2020). *Op. cit.*
[608] *Ibid.*
[609] *Ibid.*
[610] *Ibid.*
[611] *Ibid.*

implementation of Suihaoban and e-gov cloud has streamlined and created an efficient and reliable framework to digitally manage social security and financial equality. By leveraging household financial data and social security card information, the platform effectively conducts status checks on economic statuses, reducing tax fraud and increasing tax collection rates. Challenges, such as the lack of inter-department data-sharing mechanisms and high operation costs, were encountered initially but were mitigated through the introduction of Huawei's cloud computing unified platform service. Moving forward, further integration of next-generation technologies and expansion of data sharing between provincial and municipal platforms are planned to enhance government service levels and efficiency.

The Regional Health Platform has increased accessibility to healthcare services by providing online access to major hospitals and consolidating medical records. Additionally, online health platforms offer a cost-effective alternative to traditional outpatient care, facilitating health education and awareness. But the new online platform makes access to medical services technology-dependent. These apps are more commonly used by residents with higher income and higher levels of education and younger adults, while older people and those living in rural areas have expressed access difficulty issues. Challenges remain, including the need for standardisation of electronic medical records and addressing disparities in digital access, particularly for rural and older populations due to limited access to smart devices and educational literacy. To address these challenges, policies should prioritise inclusivity, technical guidance, and user literacy enhancement to ensure equitable access.

Digital inclusion within Guangzhou's smart city framework extends to minority groups, with special attention to people with disabilities, and is supported by national policies and initiatives dating back to 2007. Major ICT companies like Alibaba, Tencent, and Baidu play pivotal roles in supporting top-down nationwide initiatives by providing accessible services and creating a supportive ecosystem for people with disabilities. This comprehensive approach has positioned Guangzhou as a leader in O2O development within China, enhancing citizens' digital engagement and social services.

CHAPTER 7

SEOUL: SMART YOUTH EMPOWERMENT

7.1 City Overview

Location[612]	Seoul, Capital City of South Korea
Land Size[613]	605.25 km^2
Climate[614]	Humid continental climate with four seasons
Geography (topology/ landscape)[615]	Mountainous and hilly terrain, roughly divided into two halves of north and south by Han River; five UNESCO World Heritage Sites within Seoul Capital Area
Environmental Risks (if any)[616]	Air pollution
Population[617]	9,975,709 (2022)

[612] World Population Review. (n.d.). *Seoul Population 2024.* https://worldpopulationreview.com/world-cities/seoul-population.
[613] *Ibid.*
[614] Weather25.Com. (n.d.). *Seoul Local Weather for Today, 14 Days, and Monthly Forecast.* https://www.weather25.com/asia/south-korea/seoul.
[615] *Ibid.*
[616] Statista. (n.d.). *South Korea: Leading Environmental Issues 2019.* https://www.statista.com/statistics/999569/south-korea-leading-environmental-issues/.
[617] World Population Review (n.d.). *Op. cit.*

GDP[618]	US$388.8 billion (2022)
GDP Per Capita[619]	US$38,975 (2022)
Key Urban Issues of Smart City[620]	Comparatively high prevalence of internet usage addiction among children and adolescents

Seoul, the capital and the largest metropolis of South Korea, is situated in north-west South Korea, adjoining neighbouring Incheon metropolis and Gyeonggi province. The City of Seoul has an area of 605.25 km^2, including 25 districts, which are called gu. The gu varies significantly in land area (from 10 km^2 to 47 km^2) and populations from fewer than 140,000 to 630,000.

Influenced by the monsoon, Seoul has a humid continental climate with four distinct seasons: The coldest month, January, has a mean temperature of −6°C, while the warmest month, August, can surpass 39.6°C with a mean temperature of about 25°C.[621] Seoul has a severe issue of air pollution. In 2014, the annual average PM 2.5 concentration was 24 μg/m^3, indicating that Seoul's pollution is 2.4 times higher than the World Health Organisation Air Quality Guidelines for the annual mean PM 2.5.[622]

In 2021, Seoul had a population of 9,985,652 (around 20% of Korea's total population) and a density of 16,000 people per square kilometre, making it one of the most densely populated cities in South Korea.[623] Seoul's population density is twice and four times higher than New York

[618] Korea Stat 통계청. (n.d.). 통계청. 통계청. https://kostat.go.kr/ansk/?bmode=read&bSeq=&aSeq=379534&pageNo=1&rowNum=10&navCount=10&currPg=&searchInfo=&sTarget=title&sTxt=.

[619] *Ibid.*

[620] HRNIC. (n.d.). 한국인터넷정보센터. *Reference.* https://krnic.or.kr/jsp/eng/policy/referenceList.jsp.

[621] Weather25.Com. (n.d.). *Op. cit.*

[622] Statista. (n.d.). *Op. cit.*

[623] Rogers, L. T. & Borsella, C. P. (2017). *Growth or Decline: Understanding How Populations Change*. Census.Gov. https://www.census.gov/newsroom/blogs/random-samplings/2016/03/growth-or-decline-understanding-how-populations-change.html.

City and Los Angeles, respectively. The population is mainly homogeneous, ethnically Korean, with a small number (3.8% in 2021 Q1) of foreign residents who are primarily from Taiwan, Japan, and China of Korean ancestry.[624]

There are, however, signs of population stagnation and decline since 2011 due to multiple reasons, such as a rapidly ageing population, high costs of living causing a continuous outflow of residents to other cities, and a low fertility rate.[625] In 2021, Seoul's fertility rate was the lowest in the world, at 0.64.[626] The share of single-person households has increased to 42.2% of total households in the city (36.3% in 2011), while youth aged 9–24 years made up 15.2% of its total population and is expected to decline with low birth rate in the long term.[627]

Seoul is governed by the Seoul Metropolitan Government (executive body) and Seoul Metropolitan Council (legislative body). There are three tiers of administration: si (city), gu (district), and dong (neighbourhood).[628] As the capital city of South Korea, Seoul also deals with the nation's administrative affairs, such as correctional institutions, public education, libraries, public safety, recreational facilities, sanitation, water supply, and welfare services.[629] The Seoul Metropolitan Government has one mayor and three vice-mayors, one in charge of political affairs and the other two overseeing administrative affairs. The mayor and members of the metropolitan government are elected to 4-year terms.

Seoul is the economic, political, and cultural hub of South Korea. Seoul is the country's financial centre, and the largest proportion of the

---

[624] Eun-byel, I. (28 February 2024). South Korea's Fertility Rate Drops to New Low. *The Korea Herald*. https://www.koreaherald.com/view.php?ud=20240228050718.

[625] OECD ilibrary. (2019). *The Governance of Land Use in Korea: Urban Regeneration*. https://www.oecd-ilibrary.org/sites/ded0e64c-en/index.html?itemId=/content/component/ded0e64c-en; Rogers & Borsella (2017). *Op. cit.*

[626] Eun-byel (28 February 2024). *Op. cit.*

[627] Seoul Solution. (2021). *The Statistics of Seoul*. https://www.seoulsolution.kr/en/content/statistic-seoul.

[628] *Ibid.*

[629] Official Website of The Seoul Metropolitan Government. (n.d.). *Organization Chart*. https://english.seoul.go.kr/city-hall/organization-chart/1-seoul-metropolitan-government/.

city's workforce (70%) is employed in the services sector.[630] It is also the country's primary tourist destination with its concentration of amenities and historical sites (UNESCO World Heritage sites) as well as cultural production and pop-culture of Hallyu, or the Korean Wave where Korean music and film have become an internationally known genre.[631] The city generated 22% of the country's GDP, US$408,265 million in total, and US$38,489 GDP per capita in 2019.[632] Reflecting the economy of South Korea, Seoul has a highly developed mixed economy dominated by family-owned conglomerates (e.g., Samsung, Hyundai, and LG, which operate in the automobile, manufacturing, and electronics industries). Manufacturing is a critical sector in the city where information technology and electronics industries have replaced traditional textiles, clothing, machinery, and chemicals manufacturing.

Known as a technologically advanced city, Seoul has become the world's most wired city in 2010.[633] A high 99% of Korean households are connected to the internet and over 90% of Seoul citizens are smart device users. The city's technology infrastructure includes 20,000 internet cafes and subway Wi-Fi. South Korea is the first country in the world to commercialise 5G communications, and Seoul had the vision of becoming a "free data city" by 2022 where anyone could use Wi-Fi free of charge anywhere in the city. The city government is actively harnessing big data to provide public services, including an integrated public transport service and a digital communication system that connects smart technology to citizens and enhances citizen engagement, e.g., e-government system, e-civil complaints, and e-voting. The ubiquitous, super-fast internet connectivity has facilitated the digitalisation of Seoul and the development of smart neighbourhoods, streets, and urban interventions. According to the IMD-SUTD Smart City Index in 2023, Seoul ranked 16th in position

[630] Shin, K. H. & Timberlake, M. (2020). *Seoul, Korea's Global City: A New Urbanism for Upward Mobility* (1st edn.). Routledge. https://doi.org/10.4324/9781315122298.

[631] World Cities Culture Forum. (n.d.). *Seoul: Profile, Data, Infrastructure, Trends.* http://www.worldcitiescultureforum.com/cities/seoul.

[632] Korea Stat 통계청 (n.d.).*Op. cit.*

[633] Lankov, A. (30 January 2011). How Korea Has Become World's Most Wired Country. *The Korea Times*. https://www.koreatimes.co.kr/www/nation/2024/04/113_80600.html.

place, advancing from 23rd in 2019, 20th in 2020, and 18th in 2021.[634] The city aims to move to a knowledge-based and service economy, transforming Seoul into a liveable megacity with artificial intelligence, smart mobility, and green spaces.[635]

7.2 Smart City Development: From U-City to Smart Sustainability

South Korea started to develop a smart city in the form of U-City (ubiquitous-city) in 2001, addressing greenfield development to improve living standards.[636] It enacted the U-Cities Act in 2008 (the world's first smart city act) and established an institutional framework in 2012 to formulate U-City plans, revitalise their services, and develop new technologies. The U-City development, however, did not gain momentum due to several reasons, including reliance on developers for investment, lack of continuous innovation, and fragmented approach.[637] In 2017, the Smart City Act was passed, and the term ubiquitous was changed to smart to help improve public understanding of the development concept.

The smart city strategy is recast to be human-oriented (not just technology-driven) with a focus on a converged and linked city (instead of fragmented policy and projects) and the city as an engine for innovation-based growth (not just urban development). It aims to improve the quality of life for citizens, enhance the sustainability of cities, and foster new industries by utilising the innovative technologies of the Fourth Industrial

---

[634] IMD & SUTD. (2023). *Smart City Index 2023.* https://www.imd.org/wp-content/uploads/2023/04/smartcityindex-2023-v7.pdf.

[635] Lim, Y., Edelenbos, J., & Gianoli, A. (2023). Dynamics in the governance of smart cities: Insights from South Korean smart cities. *International Journal of Urban Sciences, 27*(sup1), 183–205.

[636] Leem, Y., Han, H., & Lee, S. H. (2019). Sejong smart city: On the road to be a city of the future. In S. Geertman, Q. Zhan, A. Allan, & C. Pettit (Eds.), *Computational Urban Planning and Management for Smart Cities* (pp. 17–33). Springer International Publishing. https://doi.org/10.1007/978-3-030-19424-6_2.

[637] World Bank. (2022). *Smart Cities in the Republic of Korea.* https://documents1.worldbank.org/curated/en/099501509212220541/pdf/IDU09bc4586900d9a047080a9de0afa91af324a9.pdf.

Revolution.[638] At the national level, a special committee on smart cities and a government committee on a smart city were set up. Three types of smart city projects were identified in cities: a new city under the national pilot smart city, a growing city under the smart city challenge, and a shrinking city under city regeneration.

Under the leadership of Mayor Park Won-Soon, since his election in 2011, Seoul has accelerated smart city development, leveraging technology to improve residents' lives under Smart Seoul 2015. Smart Seoul 2015 plans to build up smart infrastructure from 2011 to 2012 based on the existing ICT projects, providing smart services over the next 2 years (2013–2014) and advancing smart services in 2015. Smart Seoul 2015 highlights three pillars to achieve Smart Seoul[639]:

- Develop ICT infrastructure to provide smart city services.
- Have an integrated city-management framework that works in harmony with the strict adherence to common standards.
- Promote smart-tech users by increasing access to smart devices and education on their use across a diverse range of people.

Smart Seoul 2015 further seeks to enhance Seoul's reputation as a global ICT leader by increasing its sustainability and competitiveness through smart technologies and a people-oriented approach in four areas: citizen-oriented digital city, digital economy, digital social innovation, and global digital leadership.[640] Key components of Seoul's smart city master plan include the following[641]:

- Smart services for an open data approach that makes sharing government data a default.
- Smart citizens via online engagement to promote active citizen interaction.

---

[638] Smart City Korea. (n.d.). *What is a Smart City?* https://smartcity.go.kr/en/소개/.

[639] ITU-T Technology Watch Report (2013). *Smart Cities Seoul: A Case Study.* https://www.itu.int/dms_pub/itu-t/oth/23/01/T23010000190001PDFE.pdf.

[640] Ibid; Smith, L. (29 August 2018). *Seoul Smart City Portrait: Collaboration and Smart Solutions — Part 2.* Bee Smart City. https://www.beesmart.city/en/smart-city-blog/seoul-smart-city-part-2-collaboration-and-smart-solutions.

[641] Smart Cities. (2021). *Seoul.* https://www.aboutsmartcities.com/smart-city-seoul/.

- Smart infrastructure through city-wide public IoT network.
- Systematic mining of big data to improve municipal services.

Seoul's smart city development plays a significant role in achieving sustainability by promoting efficient and sustainable use of resources, reducing greenhouse gas emissions, and enhancing the quality of life for residents. For instance, Digital Youth Empowerment is a crucial component in Seoul's smart city development towards sustainability. It equips young people with the digital skills, knowledge, and tools they need to develop sustainable solutions and become active citizens in shaping the city's future.

Seoul Smart City Timeline

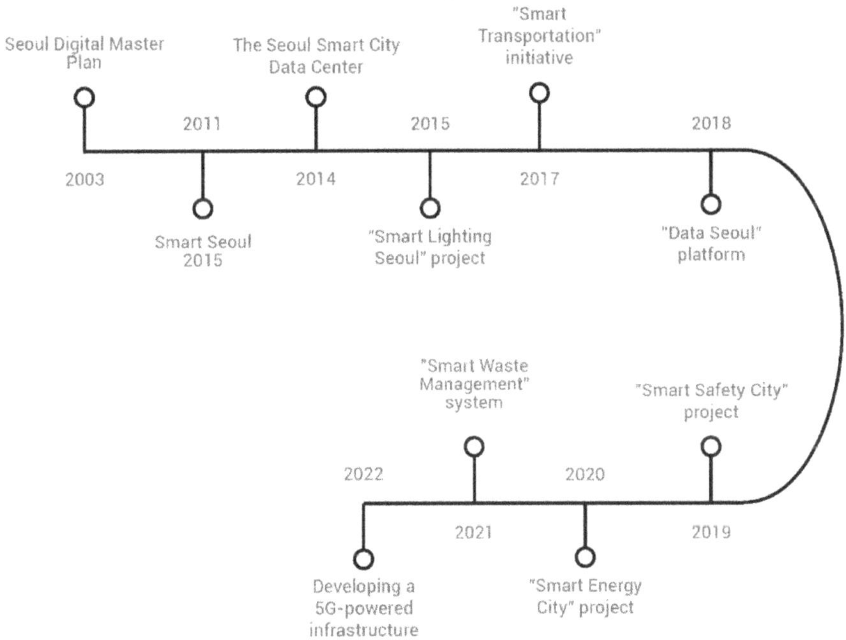

7.3 Digital Youth Empowerment

Digital youth empowerment is essential for Seoul's smart city development towards sustainability since it engages and empowers young people to contribute to sustainability efforts, fosters innovation and collaboration, and promotes social inclusion and equity, all crucial to achieving the UN sustainable development goals. Specifically, by using digital technologies and platforms, youth can participate in decision-making processes, develop innovative solutions, and drive more significant social and environmental changes.

However, studies have shown that South Korea, including its capital city of Seoul, has a comparatively high prevalence of internet addiction among children and adolescents. According to the latest report by the Ministry of Science and ICT and Korea Internet Security Agency, the internet usage rate for children aged 3–9 years was 83.9% in 2017, while the usage rate for teenagers is 99.9%.[642] The 2017 Internet White Paper warned that 59.5% of children and teenagers are at risk of internet addiction due to their parents being addicted.[643]

Among adolescents, it is known that prolonged use of the internet can cause social isolation and mental health problems, such as depression, change in character like aggression, and unhealthy behaviours, such as lack of exercise and low academic performance.[644]A study on individual's reality and ethical judgments indicates a positive connection between internet addiction and a decline in one's social intelligence.[645] The extensive exposure and dependence on the internet can also make children vulnerable to cyberbullying and other forms of online harm, which can further produce detrimental outcomes, such as depression, anxiety, loneliness, or substance abuse.[646] Adolescents and children who experienced depression during their younger days are more likely to suffer from depression in adulthood. Importantly, depression is prevalently linked to suicide, necessitating early prevention among young children.[647]

---

[642] *Ibid.*

[643] HRNIC (n.d.). *Op. cit.*

[644] Yi, Yunjeong, Hyun, Sook-Jung, Lee, Jinhwa, & An, Ji-Yeon. (2017). The impact of internet addiction on depression among Korean adolescent middle and high school students. *International Journal of Contents, 13*(2), 6–13.

[645] Park, M. H., Park, E. J., Choi, J., Chai, S., Lee, J. H., Lee, C., & Kim, D. J. (2011). Preliminary study of Internet addiction and cognitive function in adolescents based on IQ tests. *Psychiatry Research, 190*(2–3), 275–281.

[646] Kim, H. S. (2017). The impact of online social networking on adolescent psychological well-being (WB): A population-level analysis of Korean school-aged children. *International Journal of Adolescence and Youth, 22*(3), 364–376.

[647] Park *et al.* (2011). *Op. cit.*

In 2018, the Ministry of Health and Welfare (MOHW) revealed alarming findings from two weeks of monitoring suicide-related internet videos and images: 84% of 8,039 internet videos and images are related to self-mutilation and a massive increase of 3728% in such content since 2017.[648] There were extreme cases where an angry adolescent killed his parents because they attempted to correct his internet overuse habit. Similarly, there was a case where a young couple starved their baby to death while they were tending to their virtual child on the internet.[649]

The negative impacts of internet dependency and excessive internet use for the Korean community are significant and multi-faceted.[650] The Seoul Metropolitan Government took the initiative and established I Will Centres and Jump Up Internet Rescue boot camp to prevent internet addiction and to provide recovery counselling.[651] Globally, South Korea is the first country to develop a national policy and allocate a budget to tackle the high prevalence of gaming and internet-related issues.[652] This effort significantly contributed to addressing the prevalent social issue of internet addiction among children and youths in Seoul.

In 2000, the government began taking measures, such as internet addiction measurements, introducing internet prevention and counselling programmes, and conducting research studies. In 2009, the Korean government implemented an annual record survey of internet addiction for the target grades of four, seven, and ten. This is a comprehensive screening system designed to detect those at risk of developing an internet addiction.[653] The government also imposed various regulations, such as

---

[648] Lee *et al.* (2019). *Op. cit.*

[649] Seoul Solution (2014). *Op. cit.*

[650] Sung, J., Lee, J., Noh, H. M., Park, Y. S., & Ahn, E. J. (2013). Associations between the risk of Internet addiction and problem behaviors among Korean adolescents. *Korean Journal of Family Medicine*, 34, 115–122.

[651] Seoul Solution. (2014). *Op. cit.*

[652] OECD. (2018). *Children & Young People's Mental Health in the Digital Age: Shaping the Future.* https://www.oecd.org/els/health-systems/Children-and-Young-People-Mental-Health-in-the-Digital-Age.pdf.

[653] Cho, E. (2015). Therapeutic interventions for treatment of adolescent internet addiction — Experiences from South Korea. In C. Montag & M. Reuter (Eds.), *Internet Addiction: Neuroscientific Approaches and Therapeutical Interventions* (pp. 167–181). Springer International Publishing. https://doi.org/10.1007/978-3-319-07242-5_11.

banning late-night entry of minors to internet cafes and mandating the country's mobile telecommunication service providers to pre-install filtering apps against harmful content on smartphones that are sold to minors.[654] At the same time, the administration introduced the Shutdown Law in 2012, which blocks gamers under 16 years old from playing between midnight and six in the morning. However, it has little effect since they can bypass the restrictions using other adult accounts.[655]

To meet the immense amount of counselling requests, the government offered professional training in internet addiction counselling through government institutions, such as the Korea Youth Counselling and Welfare Institute (KYCI) and the National Information Society Agency (NIA).[656] These two institutions developed individual counselling protocols and group counselling programmes for each age group. They disseminated a programme manual for frontline counsellors to use and to improve the quality of services. Specifically for Seoul, the Seoul Metropolitan Office of Education issued a guideline in 2010 requesting the designation of an information ethics teacher in each school and implemented internet addiction prevention education.[657]

The Ministry of Gender Equality and Family, through the ministry's census of students, provided the city with the data for youths listed as being in danger of addiction. Hospitals in Seoul are also allowed to provide diagnoses and treatments for children and adolescents who have symptoms of depression. Despite the economic recession and budget constraints, the Seoul Metropolitan Government allocated a significant amount of its budget to prevent internet addiction in the city. In 2012, the city allocated US$2.5 million to tackle this issue.[658] The amount was increased the following year by adding another I Will Centre, demonstrating the importance and urgency of the issue for the city.

---

[654] Open Net (Korea). (15 April 2015). *Warning: Minor Smartphone Spying Law Now Comes into Force.* https://www.opennetkorea.org/en/wp/1248.

[655] Kim, D. & Shin, E. (9 January 2013). Horses to the Rescue of Korea's Internet-addicted Teens. *Reuters.* https://www.reuters.com/article/idUSDEE90800L/.

[656] Cho (2015). *Op. cit.*

[657] Seoul Solution (2014). *Op. cit.*

[658] Urban Sustainability Exchange. (n.d.). *"I Will Centers" — Internet Addiction Prevention & Intervention Centers for Children and Youths.* https://use.metropolis.org/case-studies/seoul-i-will-centre.

In 2017, seven South Korean government departments, such as the Ministry of Science, ICT and Future Planning (MSIP), and the Ministry of Gender Equality and Family (MOGEF), collaborated to address internet addiction. The budget for the internet addiction prevention programme was also increased by 55%; US$43 million of the allocated budget was spent on various measures against internet addiction, including home visit counselling services. Another US$40 million was used to diagnose children at risk of internet addiction and run a separate camp programme for addiction treatment. The Ministry of National Defence (MND), Ministry of Justice (MOJ), and Ministry of Public Safety and Security (MPSS) were taking non-budgetary actions to stop the spread of internet addiction through prevention education programmes for military personnel and juveniles on probation.[659]

7.4 Implementation and Impact of Digital Youth Empowerment

South Korea has sought to transform itself into a healthy and safe place for youth regarding internet use, introducing different initiatives that seek to make the internet a safe space for young individuals. This section elaborates on the implementation and impact of two of these innovations: the I Will Centre and the Jump Up Internet Rescue boot camp that has recorded significant achievements in recovery counselling and preventing addiction since 2007.

7.4.1 *The I Will Centre*

The I Will Centre in Seoul is a notable example of an initiative contributing to digital youth empowerment by providing digital education and training opportunities. The I Will Centre is an internet addiction prevention counselling centre that was established in 2007 with the vision to tackle the social issue of internet addiction and help children and

[659] The Korea Bizwire (14 March 2017). *S. Korean Government to Pump $13 Million into Preventing Smartphone and Internet Addiction.* http://koreabizwire.com/s-korean-government-to-pump-13-million-into-preventing-smartphone-and-internet-addiction/77419.

adolescents in Seoul lead a healthy lifestyle.[660] The initiative aims to equip young people with the skills and knowledge necessary to thrive in a rapidly evolving digital economy, while also promoting innovation and entrepreneurship among young people.

The I Will Centre offers a range of programmes, including coding camps, robotics classes, and entrepreneurship workshops. These programmes are designed to provide young people with practical, hands-on experience in key areas of digital literacy, such as coding, programming, and data analysis. The centre has partnered with local businesses and organizations to provide young people with real-world experience and mentorship opportunities, which help bridge the gap between classroom learning and real-world practice. In addition to providing training and mentorship opportunities, the I Will Centre serves as a hub for digital innovation and entrepreneurship in Seoul. The centre hosts regular networking events and hackathons, which bring together young people, industry professionals, and investors to collaborate and develop innovative solutions to real-world problems. By fostering innovation and entrepreneurship among young people, the I Will Centre helps create a more dynamic and resilient economy while also contributing to the overall growth and development of Seoul as a smart city.

The centre provides a range of services, such as counselling on internet addiction (in individual and group settings), addiction prevention education, research, campaigns, and advocacy. I Will Centre offers services not only to children and adolescents but also to parents. The centre targets high-risk internet-addicted middle school students upon recommendation from teachers. Specialists and experts are sent to provide in-home and school-based counselling, ensuring that individuals living in seclusion are included. Such an effort is critical since those addicted to the internet are more likely to be depressed and exclude themselves from the community.

Educational programmes are tailored and adapted based on the individual's physical and mental development stage. These educational programmes are being customised for parents and teachers as well. The counselling sessions require a small payment, ranging between US$5 and

---

[660] Seoul Solution (2014). *Op. cit.*

US$30 per session. For instance, I Will Centre provides a programme, namely, Dream Tree, an integrative long-term therapy programme focusing on individuals with internet addiction due to unfavourable family circumstances. It is helpful to understand the root cause of internet addiction. Students from underprivileged families are engaged in various entertaining activities instead of sitting alone at home and spending time on the internet without a purpose.[661] To ensure the centre is well equipped with resources and materials, I Will Centre also carries out research studies and hosts academic conferences on internet addiction, prevention, and recovery.[662] Advocacy campaigns on healthy cyberculture are held throughout the year.

Five years after its launch, Seoul has established five I Will Centres across the city. Each I Will Centre is staffed with 10 full-time employees and more than 50 professional counsellors and prevention experts, based on 1-year contracts. The centres adopted this approach to ensure they could provide high-quality services at lower costs. The specialists and experts all have master's degrees or higher qualifications. They also need to complete special training about internet addiction. The I Will Centres have achieved various accomplishments through their internet addiction counselling and prevention efforts.

In 2007, 5,900 counselling and preventive services were offered, and in 2012, this was increased to 749,000, almost a 99.2% growth in 5 years.[663] The centres also attended to a total of 1.7 million cases between 2007 and 2012. Through their school screening survey, the internet addiction percentage among school students in Seoul's primary and high schools was 10% in 2009. It dropped to 6.9% in 2010 and declined to 5.4% in 2011, a positive reflection of the efforts accomplished by the centres. Initiatives like the I Will Centre play a critical role in promoting digital youth empowerment, which is essential for building a sustainable and inclusive digital economy. By providing young people with the skills,

---

[661] Cho, H., Kwon, M., Choi, J.-H., Lee, S.-K., Choi, J. S., Choi, S.-W., & Kim, D.-J. (2014). Development of the Internet addiction scale based on the Internet Gaming Disorder criteria suggested in DSM-5. *Addictive Behaviors*, *39*(9), 1361–1366.

[662] Seoul Solution (2014). *Op. cit.*

[663] *Ibid.*

knowledge, and experience they need to succeed in the digital age, such initiatives help create a more resilient and adaptable workforce while also promoting social and economic development in the broader community.

7.4.2 *Jump Up Internet Rescue Boot Camp*

The Jump Up Internet Rescue boot camp is another notable example of an initiative contributing to digital youth empowerment by providing practical cybersecurity and digital privacy education. The boot camp aims to empower young people with the knowledge and skills necessary to stay safe online and to protect their personal information in the digital age.

Introduced by KYCI in 2007, the boot camp is designed to assist internet-addicted children and adolescents.[664] KYCI's Jump Up Internet Rescue School is known to be the most successful residential programme in Korea.[665] The boot camp is a government-sponsored programme and the first in South Korea. The camp is being held in one of the forested areas in the south of Seoul. The programme employs an ecological perspective integrated with bio-psycho-social aspects, motivational enhancement therapy, behavioural therapy, cognitive behavioural therapy, and a solution-focused approach.[666] The boot camp covers various topics, including online privacy, social media safety, and cybersecurity threats. Participants in the boot camp learn how to identify and respond to online threats and how to safeguard their personal information from hackers and other cybercriminals. Through a combination of lectures, workshops, and practical exercises, participants gain a comprehensive understanding of the digital landscape and its risks and opportunities. In addition to practical education, the Jump Up Internet Rescue boot camp serves as a hub for networking and collaboration in cybersecurity. The boot camp brings together young people, industry professionals, and policymakers to share knowledge and expertise and develop innovative solutions to pressing cybersecurity challenges.

---

[664] Lee *et al.* (2019). *Op. cit.*

[665] Koo, C., Wati, Y., Lee, C.C., Ho, H.Y. (2011). Internet-addicted kids and South Korean government efforts: Boot-camp case. *Cyberpsychology, Behavior, and Social Networking, 14*(6), 391–394.

[666] Cho (2017). *Op. cit.*

Figure 7.1: South Korean children participating in Jump Up Internet Rescue School boot camp.
Source: Fackler (2007).[667]

Being physically out in the forest is part of its 12-day internet addiction intervention programme. Those who participate in the camp cannot access a computer and are only allowed an hour with their mobile phone, mainly to call their family. They would also follow a rigorous regimen of physical and group activities, such as horseback riding, building an emotional and physical connection to the real world[668] (Figure 7.1). Exposing participants to the actual physical space enables them to weaken the virtual one to which they are addicted.

This programme is voluntary. However, a two- to three-month screening process is implemented before the main programme starts. Adolescents with comorbid severe symptoms that could hinder programme participation or trouble group dynamics are screened out. Three parental visits to the camp during the 12 days are mandatory. They are also expected to attend compulsory one-on-one, group, and family counselling sessions. A mentor is matched to each participant during the whole 12 days. Mentors support and encourage the children's participation in the programme. After graduation, participants will undertake three months of follow-up

[667] Fackler, M. (18 November 2007). In Korea, a Boot Camp Cure for Web Obsession. *The New York Times*. https://www.nytimes.com/2007/11/18/technology/18rehab.html.
[668] Fackler, M. (26 November 2007). In Korea, a Boot Camp Cure for Web Obsession. *The Sydney Morning Herald*. https://www.smh.com.au/technology/in-korea-a-boot-camp-cure-for-web-obsession-20071126-gdrok6.html.

programmes, such as mentoring services and regular parental education sessions, to help parents sustain the change in their children.

The notion of boot camps is appealing as Korean society practises social order, conformity, and respect for authority.[669] Research studies have found that the boot camp programme seems to offer more advantages than outpatient settings.[670] The camp director, Yoo Soon-Duk, noted that the adolescents would display negative emotions during the first few days. However, by the third day, their mood and emotions will be more positive, and they will start to enjoy hanging around individuals of their age while they do non-virtual activities.[671] In providing practical education and networking opportunities in cybersecurity, the Jump Up Internet Rescue boot camp helps build a stronger and more resilient cybersecurity workforce and is an important element of the overall development of Seoul as a smart city. As cybersecurity threats become more complex and sophisticated, the demand for skilled cybersecurity professionals will continue to grow.

There are several critical success factors in implementing digital youth empowerment including the following[672]:

- **Stakeholder collaboration**: Cooperation from government agencies, educational institutions, and youth facilities is essential to ensure successful prevention initiatives and programmes rollout. However, the centres and advocating bodies have noted that securing cooperation

---

[669] Kwee, A. W., Komoru-Venovic, E., & Kwee, J. L. (January 2010). Treatment implications from etiological and diagnostic considerations of internet Addiction: Cautions with the boot camp approach. In *Proceedings of the International Conference of e-CASE*, Singapore.

[670] Lee, S.-Y., Kim, M. S., & Lee, H. K. (2019). Prevention strategies and interventions for internet use disorders due to addictive behaviors based on an integrative conceptual model. *Current Addiction Reports*, *6*(3), 303–312.

[671] Jeong, S. (21 October 2019). The South Korean Teens So Addicted to Phones They're Going Into Detox. *CNN*. https://edition.cnn.com/2019/10/20/asia/smartphone-addiction-camp-intl-hnk-scli/index.html.

[672] City of Seoul (4 October 2022). Interview with City of Seoul [Personal communication]; I Will Centre (28 October 2022). Interview with I Will Centre [Personal communication].

between government bodies has sometimes been challenging due to respective objectives, priorities, and interests.[673]

- **Strong leadership**: Mayor Park's tenure (2011–2020) was crucial in transforming Seoul into a smart city. It showcased strong leadership and a strategic vision that blended top-down governance with bottom-up engagement to effectively meet societal challenges.[674]
- **Focusing on fun and self-control**: Designing and implementing fun activities highlighting visualisation and stimulation, such as TikTok videos and physical services, motivates youth to actively engage, learn, and take the initiative in youth business. Understanding that digital devices are just tools for problem-solving and that one should have a strong sense of self-control instead of avoidance is critical for digital literacy training.
- **Holistic approach**: For instance, creating a recovery environment at home, including connecting with parents and grandparents, is also essential to the effectiveness of educational initiatives.

Despite the government's early detection of social issues, both Seoul and South Korea face various challenges in implementation including the following:

- **The evolving nature of the internet**: The internet is dynamic and moving at an accelerated pace even as initiatives are being rolled out to tackle it. To ensure prevention and help are effective, it is important for the programme to be evidence-based and rolled out to the right people faster than the internet since it is consistently evolving.
- **Lack of research and knowledge on internet addiction**[675]: The lack of information, especially at the start, resulted in initiatives running on a limited experimental approach. Consequently, time, effort, and resources were not utilised properly. Updated information and knowledge from surveys and research are needed for organisations to plan

[673] Seoul Solution (2014). *Op. cit.*

[674] Joo, Y.-M. (2023). Developmentalist smart cities? The cases of Singapore and Seoul. *International Journal of Urban Sciences*, *27*(sup1), 164–182.

[675] *Ibid.*

better programmes. The Seoul Metropolitan Government has realised this and has since expanded the I Will Centres, where they allocate a portion of their resources to research. More could be done in terms of knowledge exchange and peer-to-peer learning rather than each centre operating solely on its own.

- **Reluctance to participate**: Individuals who are addicted to the internet often tend to deny their addiction and may not realise there is a problem.[676] As such, they rarely seek help. Those invited to participate, either through schools or other ways, often cite being tired of coming to the centre to receive help. Those who have yet to be addicted but are at potential risk often say that addiction prevention programmes are not as interesting and exciting as the internet world. Equally, families of those who are addicted to the internet are found to avoid dealing with the core issue due to the negative association with the word 'addiction', which is often perceived as shameful.[677] This poses a problem for the government and organisations when rolling out the initiatives because they need to be one step ahead in combatting ever-evolving internet advancements while at the same time, tackling social stigma and convincing reluctant individuals to participate.

The pandemic is noted to aggravate internet addiction among younger Koreans. Not only does it increase the risk but it also challenges the accessibility of therapy and counselling services. The COVID-19 regulations, such as school closures, quarantine, and social distancing, further limit an addicted individual from seeking help outside their house environment.[678] During school closures, adolescents had to use the internet to take online classes, leading some to spend more time playing internet games instead

[676] Best, P., Manktelow, R., & Taylor, B. (2014). Online communication, social media and adolescent wellbeing: A systematic narrative review. *Children and Youth Services Review*, *41*, 27–36.

[677] Kwee *et al.* (2021). *Op. cit.*

[678] Kim, D. & Lee, J. (2021). Addictive internet gaming usage among Korean adolescents before and after the outbreak of the COVID-19 pandemic: A comparison of the latent profiles in 2018 and 2020. *International Journal of Environmental Research and Public Health*, *18*(14), Article 14. https://doi.org/10.3390/ijerph18147275.

of focusing on classwork.[679] A study comparing the numbers in 2018 (pre-pandemic) and 2020 (during the pandemic) shows that the percentage of 'moderate' gamers increased by 0.7% while 'potentially at-risk' gamers increased by 4.3%.[680]

7.5 Conclusion

The City of Seoul did not wait for central government support but acted swiftly to address the emergent internet addiction problem among its young people, demonstrating how local and regional authorities can provide strong leadership and execute impactful smart city initiatives, bypassing potential delays tied to central government support. Multi-agency collaboration and co-creation play a crucial role in Seoul's digital youth empowerment. The programme brings together various stakeholders, including youth, experts, government, businesses, and civil society, to design and implement ever-evolving internet addiction-responsive initiatives. The active involvement of citizens, particularly young people, in smart city development processes enhances the relevance and effectiveness of proposed initiatives. Keeping pace with the internet's fast-changing speed is a challenge.

Digital empowerment initiatives in Seoul, such as the I Will Centre and Jump Up Internet Rescue boot camp, aim not only to combat internet addiction among youth but also to enhance digital literacy and skills, preparing the younger generation for a digitised future. According to the Organisation for Economic Co-operation and Development, digital skills are vital to economic growth and productivity.[681] Central to Seoul's development is the adoption of a holistic approach. By encompassing preventive education, tailored support services, active family and community engagement, and school-based interventions (e.g., through fun and

---

[679] Zhu, L. (2021). The psychology behind video games during COVID-19 pandemic: A case study of animal crossing: New horizons. *Human Behavior and Emerging Technologies*, *3*(1), 157–159.

[680] Kim & Lee (2021). *Op. cit.*

[681] OECD. (n.d.). *Digitalisation and Productivity*. https://www.oecd.org/economy/growth/digitalisation-productivity-and-inclusiveness/.

outdoor activities), Seoul's approach embodies a comprehensive strategy that acknowledges and addresses the multi-faceted nature of internet addiction. The programme integrates digital literacy and skills development into its core, aiming to educate young people, including those from disadvantaged backgrounds about the risks of digital addiction, and ensure they receive the necessary support to develop healthy digital habits and be better prepared to meet the challenges and opportunities of the digital age.

Digital literacy, entrepreneurship, and civic engagement among young people are assiduously promoted with the view to foster a more equitable, innovative, and resilient society. These efforts contribute towards developing a skilled workforce that is adept in cybersecurity and other digital competencies necessary for the city's global competitiveness. The success of the holistic approach in fostering digital empowerment lies in its ability to create a supportive ecosystem that promotes sustainable digital well-being while enabling youth to harness the benefits of digital technology for both personal and professional growth.

CHAPTER 8

TOKYO: SMART HEALTH SERVICES

8.1 City Overview

Location	Tokyo, Capital City of Japan
Land Size[682]	2,191 km^2
Climate	Humid subtropical climate
Geography (topology/ landscape)	Low, alluvial plains and adjacent upland hills on Honshu main island with an average elevation of 40 m
Environmental Risks (if any)	Floods, earthquakes, volcanic eruptions
Population[683]	14.04 million (December 2020)
GDP[684]	JP¥ 109, 601,589 million (US$1.05 billion[685]) (2020)

[682] Tokyo Metropolitan Government. (n.d.). *Tokyo's History, Geography and Population*. https://www.metro.tokyo.lg.jp/ENGLISH/ABOUT/HISTORY/history02.htm.

[683] Statistics of Tokyo. (n.d.). *Tokyo Statistical Yearbook*. https://www.toukei.metro.tokyo.lg.jp/tnenkan/tn-eindex.htm.

[684] 内閣府. (n.d.). 県民経済計算 (平成 23 年度 — 令和 2 年度) (2008 *SNA*、平成 27 年基準計数) <47 都道府県、16 政令指定都市分> (Cabinet Office. (n.d.). Prefectural economic accounts (FY2011 — FY2020) (2008 SNA, FY2015 base figures) <47 prefectures, 16 government-designated cities>). https://www.esri.cao.go.jp/jp/sna/data/data_list/kenmin/files/contents/main_2020.html.

[685] According to Oanda currency converter, conversion rate on 1 December, 2020: 1JPY=0.0096 USD.

GDP Per Capita[686]	JP¥ 7.806 million (US$74,967) (2020)
Key Urban Issues of Smart City	Ageing population

Greater Tokyo comprises the Tokyo Metropolis and four neighbouring prefectures: Saitama, Chiba, Kanagawa, and Yamanashi. This chapter focuses on Tokyo Metropolis (Tokyo hereafter), Japan's capital city. Located on the main island of Honshu, Tokyo covers a land area of 2,191 km^2, encompassing 23 wards that make up the historic city of Tokyo and the neighbouring cities of Kawasaki, Yokohama, Chiba, and Saitama. Tokyo has a humid subtropical climate characterised by warm and wet summers and mild winters. Like the rest of Japan, Tokyo is influenced by monsoon circulation, with cold waves in winter and humid currents in summer. Throughout 2021, the city's average temperature is 16.5°C, with the lowest at 9.5°C in January and the highest at 30.8°C in August.[687] Most of its summer and early autumn are rainy, with an average of two to three typhoons between September and October.

In 2019, Tokyo was the most populous city in Japan, with a population of 13,929,286 (12% of the country's total population)[688] and a population density of 6,168 people per square kilometre.[689] Japan has one of the world's longest life expectancies. Tokyo's life expectancy in 2015 was 81.07 years for males and 87.26 years for females.[690] The female population outnumbers males as men's death rate is higher than women's in all age groups and widens with age. Since 2012, Tokyo has experienced population decline; there are more deaths than births, and the share of older population (65 years and older) has risen to 21%, signalling a super-aged society. Estimates indicate that Tokyo's population will continue to decline by 50% in the next 90 years; by 2025, Japan's working-age

[686] 内閣府 (Cabinet Office) (n.d.). *Op. cit.*
[687] Weather2Visit. (n.d.). *Tokyo Weather in August.* https://www.weather2visit.com/asia/japan/tokyo-august.htm.
[688] Tokyo Metropolitan Government. (2019). *Tokyo Basics.* https://www.metro.tokyo.lg.jp/english/about/tokyo/documents/2019aw_3132_tokyo_basics.pdf.
[689] Tokyo Metropolitan Government (n.d.). *Op. cit.*
[690] Tokyo Updates. (2021). https://www.tokyoupdates.metro.tokyo.lg.jp/en/2021/10/29/ss_2021_16.pdf.

population is projected to contract to 59.7% of total population.[691] By 2100, almost 50% of its residents will be over 65, raising the potential risk to Tokyo's position as a global city.[692] Japan's unprecedented demographic issues — a decrease in total population, an increase in older population, and a shrinking labour force — are challenges many other countries can soon expect to face with an ageing population.

The Tokyo Metropolitan Government is one of the largest local governments in the world. It regulates 62 municipalities, of which 23 are special wards, 26 are cities, 5 towns, and 8 villages. Each of the 62 municipality units has an elected mayor and assembly who serve a 4-year term. The metropolitan government manages broader administrative work, such as water supply, sewage system, and waste management, while local municipalities are responsible for local services, such as welfare, education, and health. To address environmental issues and promote sustainable development, the Tokyo Metropolitan Government has initiated innovative measures like the world's first emission cap and trade programme in 2010.[693]

The Tokyo Metropolitan Government comprises a Governor, a Metropolitan Assembly, 13 Bureaux, 3 Public Enterprises Bureaux, and 7 Administrative Commissions. The Tokyo Metropolitan Assembly, comprising 127 directly elected members, is the legislative arm of the whole prefecture of Tokyo. It has the authority to enact, amend, and repeal metropolitan ordinances, approve the budget, certify its settlement, and elect members of the Election Administration Commission and other such bodies to ensure efficient and cohesive urban administration for the 23 special wards.[694]

---

[691] McKinsey & The Company. (2020). *The Future of Work in Japan: Accelerating Automation After COVID-19.* https://www.mckinsey.com/~/media/mckinsey/locations/asia/japan/our%20insights/future%20of%20work%20in%20japan/the-future-of-work-in-japan_v4_en.pdf.

[692] Ryall, J. (4 September 2012). Tokyo Population to Halve in the Next 90 Years. *Business Insider.* https://www.businessinsider.com/tokyo-population-to-halve-in-the-next-90-years-2012-9.

[693] United Nations. (n.d.). *Tokyo Cap-and-trade Program.* https://sdgs.un.org/partnerships/tokyo-cap-and-trade-program

[694] Tokyo Metropolitan Government. (n.d.). *Organizational Structure of TMG.* https://www.metro.tokyo.lg.jp/ENGLISH/ABOUT/STRUCTURE/structure04.htm.

Tokyo is the centre of Japanese culture, economy, education, industry, and politics. As the world's largest urban agglomeration economy, the city is a global hub for finance, investment funds, and information companies. In 2021, Tokyo has 48 of the Fortune Global 500 companies.[695] After Shanghai, Tokyo is ranked the fourth most competitive financial centre globally in 2020 and the second most competitive in Asia.[696] In the 2020 Global Financial Centres Index, Tokyo is 3rd in infrastructure, 9th in human capital, 5th under the reputational and general categories, and 5th in the Overall 2025 City Competitiveness Index.[697] Tokyo is also ranked 14th for competitive locations in fostering the fintech industry. Being second in the world and first in Asia, Tokyo has 3,700 listed companies in the fintech industry, with a total market value of US$5.3 trillion.[698]

Comparing Japan's 2019 and 2020 GDP estimates, the decrease of 28.1% in 2020 accounts for the most significant drop since the Second World War. Consumer expenditure fell sharply due to the COVID-19 pandemic. To establish a "new normal" that combines infection prevention with socio-economic activities, the Tokyo Metropolitan Government announced the initiative, Grand Reform of Tokyo 2.0, with the slogan "Determine the future of Tokyo with the people of Tokyo".[699] The establishment of a Smart Tokyo that harnesses the potential of digital technology to improve the lives of Tokyo residents while revitalising the economy has proven critical for the "new growth" of Tokyo's economy.[700]

---

[695] *CNN.* (2021). *Global 500.* https://money.cnn.com/magazines/fortune/global500/2012/countries/Japan.html.

[696] GFCI team. (2020). *The Global Financial Centres Index 28.* https://www.longfinance.net/media/documents/GFCI_28_Full_Report_2020.09.25_v1.1.pdf.

[697] Citigroup. (4 June 2013). *New Citi-commissioned EIU Report Projects Competitiveness of 120 of the World's Major Cities in 2025.* https://www.citigroup.com/global/news/press-release/2013/new-citi-commissioned-eiu-report-projects-competitiveness-of-120-of-the-world-s-major-cities-in-2025-2.

[698] FinCity Tokyo. (n.d.). *Tokyo's Financial and Economic Strength.* https://fincity.tokyo/en/strength/economic/.

[699] Tokyo Metropolitan Government. (n.d.). *Advancing Forward to Grand Reform of Tokyo 2.0 to Become a Global Destination.* https://www.metro.tokyo.lg.jp/english/governor/speeches/2020/0918/02.html.

[700] Tokyo Metropolitan Government. (n.d.). *Realizing a Tokyo that Protects the Lives of Its Citizens and Generates Earnings.* (2020). https://www.metro.tokyo.lg.jp/english/governor/speeches/2020/0918/03.html.

8.2 Smart City Development: Becoming a Human-Centred Society

Japan started to develop a smart city in the form of a smart community model of urban development. The Japan Smart Community Alliance defines a smart community as "a community where various next-generation technologies and advanced social systems are effectively integrated and utilised, including the efficient use of energy, utilisation of heat and unused energy sources, improvement of local transportation systems and transformation of the everyday lives of citizens".[701] The development of smart cities in Japan has accelerated in the aftermath of the 2011 Great East Japan Earthquake. In the face of rebuilding after a natural disaster, smart cities can utilise advanced ICT infrastructure and predictive systems to improve communication methods between multiple public agencies, such as the transportation department, emergency services, energy providers, and most importantly, with citizens.[702] With planning projections indicating a 70% possibility of a major earthquake striking Tokyo within the next 30 years, smart city initiatives are expected to accelerate crisis planning, coordination preparedness, and recovery.[703]

Since 2002, Tokyo has developed smart city plans to establish smart projects. Most smart city initiatives aim to foster energy security and efficiency, increase local development in social and economic terms, and enhance regional and local competitiveness. For example, a Smart City Project was launched in 2009 to leverage the technologies of participating global companies like Kawasaki's energy systems and Sekisui House's smart house technology to collaboratively develop a one-stop solution that addresses the various needs of a smart city.[704] Tokyo has actively promoted

[701] Okubo, H., Shimoda, Y., Kitagawa, Y., Gondokusuma, M. I. C., Sawamura, A., & Deto, K. (2022). Smart communities in Japan: Requirements and simulation for determining index values. *Journal of Urban Management*, *11*(4), 500–518.

[702] GSMA. (2013). *Smart City Resilience*. https://www.gsma.com/iot/wp-content/uploads/2013/02/cl_SmartCities_emer_01_131.pdf.

[703] Tokyo Metropolitan Government. (2023). *Tokyo Sustainability Action 2023*. https://www.metro.tokyo.lg.jp/english/about/sustainable/index.html.

[704] Kawasaki Heavy Industries, Ltd. (2012). *Six Leading Companies Newly Join in Smart City Project Further Boosting Efforts to Deploy Japan's Homegrown Smart Cities Across the Globe*. https://global.kawasaki.com/en/corp/newsroom/news/detail/20120125_1e.html.

social cohesion and addressed social issues like population ageing. The focus is on improving infrastructure (e.g., implementing smart grids to ensure a steady, uninterrupted electricity distribution) and enhancing smart urban development to tackle challenges brought about by climate and population changes, e.g., developing Tokyo Super Eco Town and progressing towards a supersmart society of Society 5.0.

Society 5.0, drafted in the 5th Science and Technology Basic Plan by the Council for Science, Technology, and Innovation, was approved by a Cabinet Decision in January 2016.[705] Society 5.0 envisions transforming the Japanese way of life by converging cyberspace and physical space while remaining human-centric. With Society 5.0, Japan aims to become the first country in the world to achieve a human-centred society where everyone can enjoy a high quality of life by combining advanced technologies in diverse industries and social activities, fostering innovation to create new value, and developing a smart city that shifts from a technology-driven approach to a social issue-driven approach. Tokyo's articulation of Society 5.0 initiatives focuses on resolving various social challenges like healthcare for an ageing population, mobility in areas with population decline, finance, and other industries. The goal of Society 5.0 reinforces the notion of a smart platinum society that aims to improve people's lives through innovation, a "hundred-year life", and an "age-free society" where older people are encouraged to stay healthy and to

Tokyo Smart City Timeline

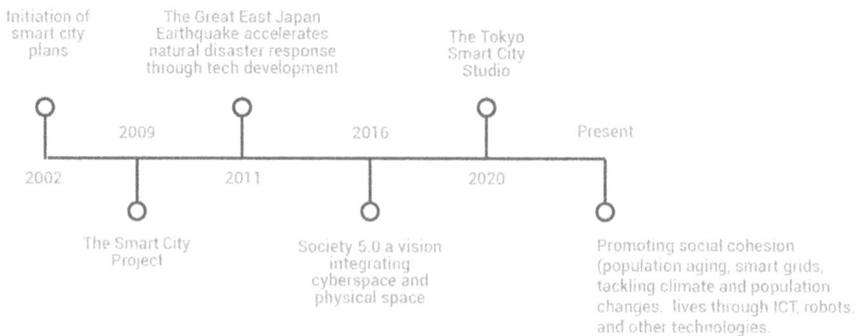

[705]Cabinet Office. (n.d.). *Society 5.0.* https://www8.cao.go.jp/cstp/english/society5_0/index.html.

continue playing active roles in the labour force and society with the use of ICT, robots, and other technologies.[706]

8.3 Digital Healthcare

Despite being one of the healthiest populations in the world,[707] Japan has been embracing healthcare information technology as a vehicle in healthcare reform since the early 2010s to address its fast-ageing population.[708] By 2065, individuals above 65 years old will account for 38.4% of the country's total population.[709] The repercussion is a super-aged society and shrinking workforce, bringing several challenges, including increased healthcare costs. The cost of care for persons aged 65 and older is three to five times more than for those under 65 years old.[710] According to the National Federation of Medical Insurance Companies, Japan's healthcare expenses will rise from US$360 billion in 2016 to US$510 billion in 2025.[711]

Against rising healthcare demands and costs for the older population, Japan's medical resources face additional strain. The Organisation for Economic Co-operation and Development (OECD) estimates that Japan's doctor number has fallen below average in 2018 (2.5 per 1,000 people).[712]

---

[706] Ministry of Foreign Affairs. (2015). *Address by Prime Minister Shinzo Abe at the Symposium of the 70th anniversary of the United Nations "What the United Nations means for Japan"*. https://www.mofa.go.jp/fp/unp_a/page3e_000310.html.

[707] World Population Rcvicw. (n.d.). *Healthiest Countries 2024*. https://worldpopulationrc-view.com/country-rankings/healthiest-countries.

[708] Abraham, C., Nishihara, E., & Akiyama, M. (2011). Transforming healthcare with information technology in Japan: A review of policy, people, and progress. *International Journal of Medical Informatics*, *80*(3), 157–170.

[709] *Kyodo News*. (2018). *Japan's Population Decline to 127 mil., Aging Continues in 2017*. https://english.kyodonews.net/news/2018/04/1fc13074ea8a-japans-population-decline-aging-continue-in-2017.html.

[710] Abraham, C., Nishihara, E., & Akiyama, M. (2011). Transforming healthcare with information technology in Japan: A review of policy, people, and progress. *International Journal of Medical Informatics*, *80*(3), 157–170.

[711] Narvaez Rojas, C., Alomia Peñafiel, G. A., Loaiza Buitrago, D. F., & Tavera Romero, C. A. (2021). Society 5.0: A Japanese concept for a superintelligent society. *Sustainability*, *13*(12), Article 12. https://doi.org/10.3390/su13126567.

[712] OECD. (n.d.). *OECD Data — Doctors*. https://data.oecd.org/healthres/doctors.htm.

Japan's healthcare workers are experiencing a workload burden. In a 2015 survey conducted by the Ministry of Health, Labour, and Welfare (MHLW), the 7th most cited reason for certified care workers to leave their care worker jobs was "low back pain".[713] Another study shows that only 14.8% of nursing-care facilities in Japan use portable patient lift devices.[714] The low utilisation is due to several reasons, such as the long time they take to use the device, the difficulty of securing slings, the risk of injuring a patient, and the patient's emotional and physical distress when using the device. Despite having one of the largest medical device markets and one of the best healthcare systems in the world, Japan is lagging in incorporating digital transformation into its healthcare system due to the conservative nature of the country's medical industry.

The government and citizens prefer affordable yet high-quality care under the country's universal health insurance system, which generally makes it less urgent to adopt advanced technologies to monitor health and prevent diseases.[715] However, with the demographic shift and introduction of Society 5.0, healthcare structure in Japan is anticipated to evolve. Under Society 5.0, there will be greater use of ICT, robots, and other such digital technologies to promote health, including improved diets, more active lifestyles, and reduced tobacco and alcohol consumption.

8.4 Implementation and Impact of Next-Generation Healthcare System and Robots

This section elaborates on the implementation and impact of two of these innovations: next-generation healthcare system (health and medical information platforms and telemedicine) and robots.

---

[713] Tsukada, N. (2021). A new era for policies for care workers in Japan: Current status and future directions. 12.

[714] Mukai, T., Hirano, S., Nakashima, H., Kato, Y., Sakaida, Y., Guo, S., & Hosoe, S. (2010). Development of a nursing-care assistant robot RIBA that can lift a human in its arms. In *2010 IEEE/RSJ International Conference on Intelligent Robots and Systems*, 5996–6001. https://doi.org/10.1109/IROS.2010.5651735.

[715] World Health Organization. (2018). *Japan health system review.* https://apps.who.int/iris/bitstream/handle/10665/259941/9789290226260-eng.pdf.

8.4.1 *Next-Generation Healthcare System*

In 2001, the Japanese government announced the "E-Japan Strategy" as a national IT policy, concentrating on digital healthcare and the promotion of IT in healthcare-related fields.[716] The E-Japan Strategy was followed by E-Japan Strategy II (in 2003), New IT Reform Strategy (in 2006), and I-Japan Strategy 2015 (in 2009). In the latest health policies, the government introduced the 2018 Growth Strategy to create the next-generation healthcare system. The main goal of the next-generation healthcare system is to lengthen life expectancy and build a next-generation healthcare system that offers services tailored to the individual's needs, to promote preventive care and improve the health of individuals. There are many initiatives under the next-generation healthcare system. This section looks at two of them: health medical platforms and telemedicine.

Both health medical platforms and telemedicine require fast and reliable ICT infrastructure. Japan is implementing 5G networks to mitigate the challenges of uneven distribution of medical resources for residents, including older population in rural parts of Japan. 5G networks are high-speed, large-capacity, low-latency wireless telecommunication networks with many projected connections to enable telemedicine, remote surgery, online pharmaceutical instruction, online collection, storage, and use of medical data and images.[717] The Ministry of Internal Affairs and Communications (MIC) is primarily responsible for establishing 5G networks in Japan.[718] In 2019, the Japanese government assigned four companies to the task, NTT DOCOMO, KDDI au, Softbank, and Rakuten Mobile, a 5G spectrum. Over the next 4 years, these four mobile carriers are expected to spend more than US$14 billion in capital expenditures to expand the 5G networks. According to Fitch Research, 4G subscriptions in Japan would decline with 5G

---

[716] Sonoda, T. (2011). Evolution of electronic medical record solutions. Fujitsu Scientific and Technical Journal, *47*(1), 19–27.

[717] Georgiou, K. E., Georgiou, E., & Satava, R. M. (2021). 5G use in healthcare: The future is present. *JSLS: Journal of the Society of Laparoscopic & Robotic Surgeons*, *25*(4), e2021.00064. https://doi.org/10.4293/JSLS.2021.00064.

[718] International Trade Administration. (2020). *Japan's 5G Networks*. https://www.trade.gov/market-intelligence/japans-5g-networks.

becoming the dominant technology by 2026 and more than 151 million 5G subscriptions in Japan by 2029.[719]

In digital healthcare, data are managed by healthcare entities and new entities, such as private firms, leading to many critical issues relating to data utilisation guidelines, personal data protection, and cybersecurity precautions. As digital healthcare advances and is gaining more collaborative efforts, information security and privacy of health, medical, and nursing care information are potential pitfalls in implementation. For the medical sector, two guidelines concerning the management of health information have been introduced[720]:

- Guidance for the appropriate handling of personal information by medical or care-related service providers.
- Guidance concerning the safety management of medical information systems.

The first guideline focuses on the information produced or recorded by healthcare practitioners.[721] According to the guideline, providers that supply medical information systems are required to receive a privacy mark or an information security management system certificate. Additionally, the second guideline outlines specific requirements for the providers to follow regarding the risk-management process (i.e., risk assessment, risk analysis, risk management, and risk communication).[722] In practice, the guidelines necessitate medical data to be stored in Japan. However, if a complete copy of the data is stored within Japan, copies of the information can be maintained outside Japan.

---

[719] Fitch Solutions. (6 August 2019). *Japan 5G: Tokyo Olympics Will Explore New Use-cases, But Smart Cities Will Drive Deployment.* https://www.fitchsolutions.com/bmi/telecommunications/japan-5g-tokyo-olympics-will-explore-new-use-cases-smart-cities-will-drive-deployment-06-08-2019.

[720] METI. (n.d.). 医療情報を取り扱う情報システム·サービスの提供事業者における安全管理ガイドライン (Safety management guidelines for providers of information systems and services that handle medical information). https://www.meti.go.jp/policy/mono_info_service/healthcare/teikyoujigyousyagl.html.

[721] DataGuidance. (1 October 2021). *Japan — Data Protection Overview.* https://www.dataguidance.com/notes/japan-data-protection-overview.

[722] DataGuidance. (20 July 2023). *Japan: MIC Publishes Safety Management Guidelines on Information Systems and Medical Information.* https://www.dataguidance.com/news/japan-mic-publishes-safety-management-guidelines.

8.4.2 *Health and Medical Information Platforms*

Health and Medical Information Platforms entail three components, namely, Electronic Health Records (EHR), Anonymous Databases (Anonymous DB), and Personal Health Records (PHR).[723] EHR is a system for linking health and medical information among various healthcare stakeholders, including medical institutions, institutions, and pharmacies, and nursing facilities, thereby enabling information to be shared electronically beyond organisational barriers.[724] Anonymous DB is mainly for public health and medical research purposes.[725] Anonymous DB is a database with which EHR or PHR information can be used for dual purposes: patient treatment for public health and medical research. PHR is a platform to share among patients and stakeholders.[726] PHR plays two primary roles. The first is to have patients understand their health and medical information, allowing them to understand their health conditions better, and promote behaviour modifications for maintaining and improving their health. The second is to share PHR information with healthcare stakeholders, enabling them to better understand the patients' health conditions and further create a higher quality of care for patients.

In March 2021, the government announced that PHR would further evolve, adding new components, such as online certification confirmation and syncing of "Mynaportal" and "My Number" cards. "Mynaportal" is an online portal operated by the Japanese government. It offers access to various services that one would usually find at their local city council office.[727] The majority of municipalities would make the online application option possible for most of the applications and procedures on the portal, including COVID-19-related support measures. However, each available service may differ from one municipality to another. "My number" card is a plastic card that contains an identification card chip and the

---

[723] Nomura Research Institute. (2020, June 10). *The Proper Role of Japan's Health and Medical Information Platforms in Light of the Current Handling of Covid-19.* https://www.nri.com/-/media/Corporate/en/Files/PDF/keyword/proposal/20200609_2_en.pdf?la=en&hash=09373C49DC1D664C491C3F11F7C62AD0CCDD872F.

[724] *Ibid.*

[725] *Ibid.*

[726] *Ibid.*

[727] Practical Japan. (2020). *7 Useful Things You Can Do on Mynaportal.* https://practicaljapan.com/what-can-you-do-on-mynaportal/.

bearer's name, address, date of birth, and sex.[728] My number card is issued to all registered residents with residential records at a municipal office. "My number" system was introduced in 2015 after the unsuccessful launch of Social Society and Tax Number System due to data privacy and security concerns.[729]

Following the improvements of the PHR, in April 2021, the Healthcare AI Platform Collaborative Innovation Partnership (HAIP) was launched.[730] The partnership was created by Nihon Unisys, Hitachi, IBM Japan, SoftBank, Mitsui, and Microsoft Japan, to establish standard fundamental technologies for a social healthcare AI platform. The current objective of the partnership includes forming functional needs, setting international standards, implementing connection APIs, and assuring security. HAIP's research findings will be made available to the public to aid the dissemination and development of healthcare AI services to ensure the quality of medical care and lessen the burden placed on medical workers.

8.4.3 *Telemedicine*

Telemedicine in Japan began in August 2015 with phone services for return-visit patients of any illness. Telemedicine or online medical care refers to a medical doctor offering patients medical care using telecommunication devices in real time, including medical examinations, conveying diagnosis results, and providing prescriptions.[731] In the past, this form of medical care was for patients continuing treatment for symptoms that have subdued after hospitalisation. However, the COVID-19 pandemic made telemedicine available to all patients, including first-time patients

---

[728] Plaza Homes. (2020). *The My Number System in Japan and How It Affects Non-Japanese.* https://www.realestate-tokyo.com/living-in-tokyo/immigration-government/my-number/.

[729] Onaya, T., Taketomi, M., Fujii, H., & Negishi, T. (2015). ICT trends in Japan's healthcare policy. *Fujitsu Scientific and Technical Journal, 51*(3), 10–17.

[730] Mitsui & Co., Ltd. (1 April 2021). *Launch of "Healthcare AI Platform Collaborative Innovation Partnership (HAIP)" Approved by MHLW and METI Ministers.* https://www.mitsui.com/jp/en/topics/2021/1240925_12171.html.

[731] COSMO. (2020). *The Acceleration of Digital Health in Japan Amid COVID-19.* https://cosmopr.co.jp/en/the-1st-cosmo-innovation-seminar-en/.

and phone-only care. According to a survey conducted by the MHLW, the number of healthcare companies using telehealth solutions in Japan increased from 14.5% in June 2020 to 15.1% in January 2021 (these numbers plateaued for the rest of the year afterwards).[732] MHLW announced that it would develop a detailed guideline to assure the treatment's safety and quality, followed by official discussions about reforming the reimbursement rate.[733]

Japan still has many challenges in the transition from its traditional healthcare structure to the next-generation digital healthcare system. The country is far behind the rest of the world in the digitisation of medical records. The dissemination of EMR systems is at 34% due to its high introductory costs and unknown cost-effectiveness.[734] Introducing an EMR system in a hospital often burdens personnel more and causes stress, especially for healthcare professionals unfamiliar with computer operations. The increased workload of doctors and other staff members may result in longer patient waiting times. Moreover, many hospitals noted that EMR adoption was expensive. While the government provided the initial funding for setting up IT systems, hospitals found that the cost of maintenance was not feasible for them in the long term.[735]

Japan's inadequacy in digital healthcare was highlighted during the COVID-19 pandemic. For instance, when the government rolled out the emergency cash handout programme during the pandemic, instead of relying on the previously launched digital 'My Number' card that cost US$8 billion, municipalities had to rely on paper and manual labour. This was because when the pandemic began, only 15% of the population had

---

[732] *APCO Worldwide.* (2021). *The Challenges and Growth of Telehealth in Japan: Is It Here to Stay?* https://apcoworldwide.com/blog/the-challenges-and-growth-of-telehealth-in-japan/.

[733] Miyatake, H., Kosaka, M., Arita, S., Tsunetoshi, C., Masunaga, H., Kotera, Y., Nishikawa, Y., Ozaki, A., & Beniya, H. (2021). Videoconferencing for home care delivery in Japan: Observational study. *Journal of Medical Internet Research, 23*(9), e23539. https://doi.org/10.2196/23539.

[734] Intralink. (2019). *Is Digital Health Finally Taking Off in Japan?* https://www.intralink-group.com/en-GB/News/Blog/April-2019/Is-digital-health-finally-taking-off-in-Japan.

[735] Yasunaga, H., Imamura, T., Yamaki, S., & Endo, H. (2008). Computerizing medical records in Japan. *International Journal of Medical Informatics, 77*(10), 708–713.

received their "My Number" card.[736] This prompted Suga, the current Prime Minister, to launch a thorough effort to go digital in a proper and orderly manner. Still an ongoing process, plans are underway to set up a digital agency to control the government's entire IT system. Suga has also resolved legal charges to enable the transfer of authority from each ministry and municipality. If these plans go as expected, they may resolve many of Japan's digitalisation challenges.

The COVID-19 pandemic has also shown that digital healthcare transformation has not been widely adopted in Japan.[737] While Japan permitted doctors to hold online consultations in April 2020, significant efforts to eliminate inefficiencies by telehealth have not received a favourable response. Among Tokyo, Osaka, and nine other prefectures whose governments implemented a state of emergency during the initial wave of the pandemic, there were only a monthly average of 2.6 online initial sessions per 100,000 people. The low numbers are due to several reasons. The Japan Medical Association is wary of telemedicine, concerned about doctors being sued for misdiagnoses or recordings from appointments being leaked.[738] Many doctors also found that online appointments entail more effort and are less helpful than in-person visits.

In sum, critical success factors for the next-generation healthcare system include the following[739]:

---

[736] Suzuki, W. (15 June 2021). COVID Gives Japan 'Last Chance' to Reverse Digital Defeat. *Nikkei Asia*. https://asia.nikkei.com/Spotlight/Asia-Insight/COVID-gives-Japan-last-chance-to-reverse-digital-defeat.

[737] Seki, Y. (19 August 2021). Telemedicine Gains No Traction in Japan Despite COVID Surge. *Nikkei Asia*. https://asia.nikkei.com/Spotlight/Coronavirus/Telemedicine-gains-no-traction-in-Japan-despite-COVID-surge.

[738] Current Status and Challenges of the Dissemination of Telemedicine in Japan After the Start of the COVID-19 Pandemic | Telemedicine and e-Health. (2021). https://www.liebertpub.com/doi/10.1089/tmj.2021.0336.

[739] Abraham, C., Nishihara, E., & Akiyama, M. (2011). Transforming healthcare with information technology in Japan: A review of policy, people, and progress. *International Journal of Medical Informatics*, *80*(3), 157–170; Ito, J., Edirippulige, S., Aono, T., & Armfield, N. R. (2017). The use of telemedicine for delivering healthcare in Japan: Systematic review of literature published in Japanese and English languages. *Journal of Telemedicine and Telecare*, *23*(10), 828–834.

- **Digital Infrastructure**: It is crucial that major telecom providers successfully deploy 5G networks, as mandated by the Ministry of Internal Affairs and Communications. These networks enable high-speed, low-latency communications essential for telemedicine and remote healthcare services, particularly in rural areas.
- **Effective data management**: As healthcare increasingly relies on digital platforms, managing and protecting personal health information become paramount. Implementing guidelines for the appropriate handling of personal information by healthcare providers and ensuring the safety management of medical information systems are critical to maintaining trust and integrity in the system.
- **Integration of health records**: The development of Health and Medical Information Platforms that include Electronic Health Records (EHR), Anonymous Databases, and Personal Health Records (PHR) is essential. These platforms must ensure interoperability among different healthcare stakeholders and facilitate seamless information sharing, which is vital for efficient healthcare delivery.
- **Stakeholder inclusion**: In particular, establishing the Healthcare AI Platform Collaborative Innovation Partnership (HAIP) suggests a strong focus on developing standard technologies for healthcare AI. This public–private partnership intends to foster innovation, set international standards, and ensure security in digital healthcare solutions.

There are several challenges in advancing the next-generation healthcare system including the following[740]:

- **High costs of technology implementation**: While the Japanese government has invested in setting up advanced digital healthcare systems, the ongoing costs and the need for continuous updates and security measures pose a financial barrier. Establishing a sustainable

[740] Nomura, S., McEnany, M., Sugawara, J., Noritake, R., & Miyata, H. (2022). *Sustainability and Resilience in Japan's Health System*. Health and Global Policy Institute. https://hgpi.org/en/wp-content/uploads/sites/2/Report_PHSSR_Japan_2022_ENG.pdf.

funding model for ongoing operations, updates, and scaling is critical but challenging.

- **Workforce training**: As traditional healthcare practices are deeply ingrained, the healthcare community, including doctors and nurses, may resist transitioning to digital systems due to concerns about efficacy, security, and reliability. Overcoming this resistance requires a cultural shift within the healthcare industry and robust training programs. Thus, substantial training and adaptation for healthcare workers to become proficient with new technologies are vital but also a financial and time-consuming endeavour.
- **Privacy concerns and public acceptance**: As digital healthcare involves handling sensitive personal information, there are substantial data privacy and security concerns. The fear of data breaches and the potential misuse of personal health information can hinder the adoption of digital health platforms. Building trust in digital health systems, particularly among older adults who may be less tech-savvy, is essential. Misinformation and scepticism about the effectiveness of digital health services can be significant barriers.

8.4.4 *Next-Generation Robots*

In June 2014, following approval from the cabinet, the Ministry of Economy, Trade, and Industry (METI) and MHLW introduced the notion of next-generation robots as device assistance in both healthcare and everyday life. The introduction of next-generation robots aims to reduce the patient care burden of healthcare professionals and nursing staff, e.g., helping with heavy lifting of patient, and lessen the challenges of labour shortage.[741]

In February 2015, as a core strategy of the Abenomics Growth Strategy, the Japanese government launched the New Robot Strategy, based on the updated 2014 Japan Revitalization Strategy.[742] The New Robot Strategy's 5-year plan (2016–2020) focused on five sectors:

[741] Prime Minister's Office of Japan. (10 February 2015). *New Robot Strategy.* https://www.kantei.go.jp/jp/singi/keizaisaisei/pdf/robot_honbun_150210EN.pdf.
[742] *Ibid.*

manufacturing sector; service fields and nursing and medical fields; infrastructure, disaster response, and construction fields; and agriculture, forestry, fishery, and fishery and food industry. The priority area for the Nursing field was to support the development and emphasise the importance of five development fields[743]:

- Transfer support from the bed.
- Walking support.
- Excretion support.
- Monitoring people with dementia.
- Bathing support.

One of the 2020 visions is to increase awareness of nursing methods using the newest robot technology. To implement these plans, the Robot Revolution and Industrial IoT Initiative (RRI) were established in May 2015.[744] The government and RRI developed a 3-stage plan for sectoral robot-related R&D projects covering the following:

- Introduction and substantiation.
- Development of technology applicable to the market.
- Development of next-generation technology.

According to the New Robot Strategy, 10 different types of robotics are available. This writing will focus on three of these: therapeutic robots, interactive robots to combat loneliness, and smart clothing that monitors well-being through the Internet of Things (IoT).

8.4.5 *Therapeutic Robots: Paro*

These are therapeutic aids robots (Figure 8.1). Across many different cognitive disorders, dementia is one of the most challenging conditions

---

[743] *Ibid.*

[744] Robot Revolution & Industrial IoT Initiative. (15 May 2015). *The Founding General Meeting of the Robot Revolution Initiative was Held.* https://www.jmfrri.gr.jp/english/info/398.html.

Figure 8.1. Demonstration of PARO's interaction with older adults.
Source: Quartz.[745]

among older adults in Japan. A 2018 study on dementia in Japan estimated that the number of people with dementia would reach 5.3 million by 2070.[746] In 2020, it was recounted that a total of 17,565 people with dementia were reported missing to the police; in Tokyo, 1,169 people were reported missing.[747] These numbers precipitated the urgency to have therapeutic robots, such as PARO, in the healthcare industry.

Paro is an interactive and therapeutic robot. The Paro project was initiated in 1993 by Japan's National Institute of Advanced Industrial Science and Technology (AIST). According to Takanori Shibata, a head

[745] Thang, L. (12 December 2018). Before We Give Them Fuzzy Robots, Let's Try Solving Elderly Loneliness with People. *Quartz*. https://qz.com/1490605/before-we-give-them-fuzzy-robots-lets-try-solving-elderly-loneliness-with-people.

[746] Fukawa, T. (2018). Prevalence of Dementia Among the Elderly Population in Japan. *Health Prim Car*, *2*(10), 15761.

[747] *The Japan Times*. (24 June 2021). *Number of Dementia Patients Reported Missing Hits Record-high 17,565 in 2020*. https://www.japantimes.co.jp/news/2021/06/24/national/dementia-missing/.

research scientist at the National Institute of Advanced Industrial Science and Technology, Paro took more than 10 years to develop and received around US$20 million in government funding.[748] The Paro project debuted the first-generation robot in 1998. Several more types were produced in the years that followed. The eighth generation of Paro was created in 2004, and the robot has been sold in Japan since 2015.[749]

Paro is shaped like a baby harp seal and weighs the same as a newborn baby. A baby harp seal design is adopted because it is deemed a more neutral animal. Paro is designed to replicate the psychological, physiological, and social benefits of animal therapy. Paro has tactile sensors, can move its tail and flippers, and open its eyes when pet. Paro can learn names and respond to sounds and words its owner uses frequently. Additionally, it produces sounds like a baby seal and is programmed to be active during the day and sleep at night, just like a typical human. The artificial intelligence software changes the robot's behaviours based on sensors that monitor sound, light, temperature, and touch. Remarkably, it can show emotions, such as surprise, happiness, and anger and will cry if it is not receiving sufficient attention.

On the market since 2005, Paro is one of the earliest robots deployed in dementia care; it is used in care facilities as therapy for dementia patients. Paro is used in the United States, Canada, Europe, Asia, and Oceania, among other developed nations and regions, to assist in delivering person-centred care.[750] Paro is an FDA-approved medical device in the United States. Awarded as the world's most therapeutic robot by Guinness World Records, Paro has been found to help alleviate stress and reduce depression and anxiety in patients with dementia by improving their relaxation and motivation. It has been featured by *The New York Times*, *Wall Street Journal*, *CNN*, and most recently on Aziz Ansari's new

---

[748] World Economic Forum. (3 April 2018). *Robots are Helping to Take Care of Japanese Senior Citizens*. https://www.weforum.org/agenda/2018/04/robots-are-the-new-workforce-for-taking-care-of-japanese-seniors/.

[749] IEEE. (n.d.). *Paro — ROBOTS: Your Guide to the World of Robotics*. https://robots.ieee.org/robots/paro/

[750] Inoue, K., Wada, K., & Shibata, T. (2021). Exploring the applicability of the robotic seal PARO to support caring for older persons with dementia within the home context. *Palliative Care and Social Practice*, *15*, 26323524211030285. https://doi.org/10.1177/26323524211030285.

Netflix series, *Master of None*.[751] Paro is the subject of clinical studies in at least ten countries, including Denmark, where local governments buy Paro for their nursing homes. A 2014 study examining the interactions between Paro and older adults with dementia in a nursing home showed that more positive emotions were found among dementia patients when interacting with Paro.[752] It is important to highlight that Paro is not developed to be a substitute for human care but more as a social icebreaker and a compassionate assistant.

In Japan, around 60% of Paro robots are sold to domestic users, with the remaining 40% sold to healthcare institutions and museums.[753] While Paro provides technology opportunities for dementia care and managing challenging behavioural symptoms, Paro adoption in care settings remains limited. The high cost and workload concerns, stigma, and ethical considerations are significant impediments to this technology adoption.[754] The price of the robot was US$6,000 in 2009,[755] considered as a barrier to its potential extensive usage in care settings.[756] Another concern is about infection prevention and control. As the fur covering is not intended to be frequently removed or machine washed, keeping Paro's fur clean can be a problem, particularly for immunocompromised patients.[757] The suggested protocol involves cleaning Paro between contact with different users to

---

[751] Vaswani, J. (10 March 2020). *This Japanese scientist created an AI therapy robot for elderly patients*. Hive Life. https://hivelife.com/paro/.

[752] Koh, I. S. & Kang, H. S. (2018). Effects of intervention using PARO on the cognition, emotion, problem behavior, and social interaction of elderly people with dementia. *Research in Community and Public Health Nursing*, *29*(3), 300–309.

[753] Šabanović, S. (2014). Inventing Japan's 'robotics culture': The repeated assembly of science, technology, and culture in social robotics. *Social Studies of Science*, *44*(3), 342–367.

[754] Hung, L., Liu, C., Woldum, E., Au-Yeung, A., Berndt, A., Wallsworth, C., Horne, N., Gregorio, M., Mann, J., & Chaudhury, H. (2019). The benefits of and barriers to using a social robot PARO in care settings: A scoping review. *BMC Geriatrics*, *19*(1), 232.

[755] PARO Therapeutic Robot. (n.d.). *Press releases*. http://www.parorobots.com/pressreleases.asp.

[756] Hung *et al.* (2019). *Op. cit.*

[757] Moyle, W., Bramble, M., Jones, C., & Murfield, J. (2018). Care staff perceptions of a social robot called Paro and a look-alike Plush Toy: A descriptive qualitative approach. *Aging & Mental Health*, *22*(3), 330–335.

prevent germs from spreading, potentially adding an additional workload for care setting staff. Another issue raised by caregivers and workers was the stigma associated with interacting with a robot animal, creating risks of infantilising and dehumanising care.[758]

8.4.6 *Robots to Combat Loneliness: PaPeRo*

Another next-generation robotic is the interactive type of robot (Figure 8.2). Depending on their functions, these robots use verbal and non-verbal communication to combat loneliness and enable older adults to live independently. Studies have shown that the use of robot technology in one's everyday life can be effective in curbing loneliness among older adults.[759] Contrary to other kinds of robots, these robots are primarily personal.

PaPeRo, which stands for "Partner-type-Personal-Robot", is a personal robot developed by NEC Corporation in 2001.[760] Supported by the Japan Science and Technology Agency under the Strategic Promotion of Innovative Research and Development Programme, PaPeRo is adopted as a platform where older adults, welfare and homecare service workers, and engineers collaborate to design and build an interactive information support system for older adults, mainly persons with dementia. This robotic project aims to establish an environment where the older population can enjoy living each day with dignity by utilising Information and Robot Technology (IRT) and Information and Communication Technology (ICT).[761] The R&D project was conducted based on field-based innovation, with PaPcRo tested by older adults.

---

[758] Moyle, W., Bramble, M., Jones, C. J., & Murfield, J. E. (2019). "She had a smile on her face as wide as the great australian bite": A qualitative examination of family perceptions of a therapeutic robot and a plush toy. *The Gerontologist*, *59*(1), 177–185.

[759] van der Heide, L. A., Willems, C. G., Spreeuwenberg, M. D., Rietman, J., & de Witte, L. P. (2012). Implementation of CareTV in care for the elderly: The effects on feelings of loneliness and safety and future challenges. *Technology and Disability*, *24*(4), 283–291.

[760] Owano, N. (13 November 2013). NEC Introduces the PaPeRo Petit Robot. *Phys Org*. https://phys.org/news/2013-11-nec-papero-petit-robot.html.

[761] Research Institute of the National Rehabilitation Center for Persons with Disabilities. (n.d.). *An Informative and Interactive Robot to Help in Your Daily Life*. http://www.rehab.go.jp/ri/kaihatsu/papero_html_e/index.html.

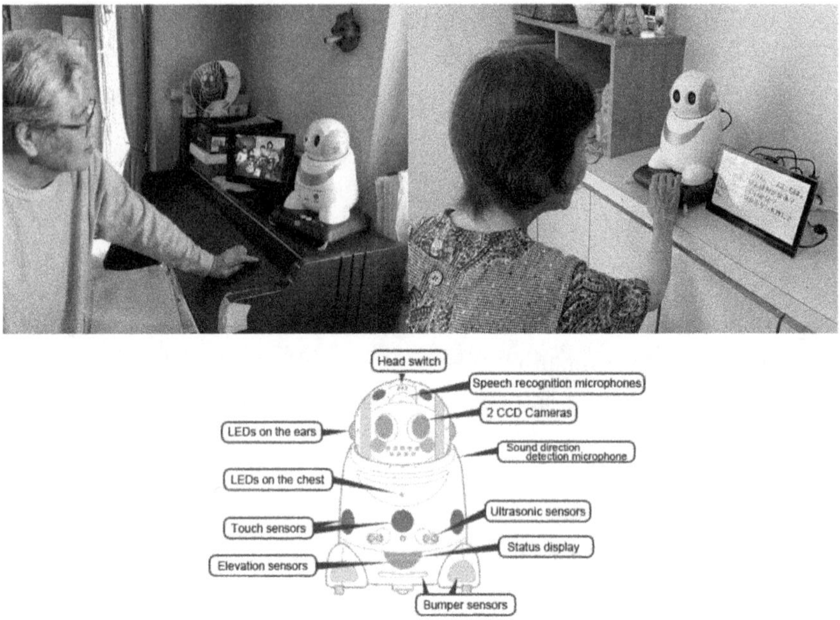

Figure 8.2. PaPeRo's interaction with older adults.
Sources: Nikkei,[762] NEC,[763] and RazarRobotics.[764]

PaPeRo was first field-tested at EXPO 2005 in Aichi, Japan.[765] Over the years, NEC has produced different PaPeRo variants, but their interactive function remains the same. Designed for household use, PaPeRo is not

---

[762] Nikkei. (23 April 2018). *NECが「PaPeRo i」で高齢者見守り、愛媛県西条市で実証* (Monitoring older adults, demonstrated in Saijo City, Ehime Prefecture). https://xtech.nikkei.com/dm/atcl/news/16/042211031/.

[763] NEC. (4 August 2020). *NEC、高齢者見守りにおけるCOVID-19対策としてコミュニケーション·ロボットを活用したサービス「みまもり　パペロ」を福井県坂井市へ提供* (NEC provides "Mimamori Papero," a service that uses communication robots to monitor older adults as a COVID-19 countermeasure, to Sakai City, Fukui Prefecture). https://jpn.nec.com/press/202008/20200804_01.html.

[764] Razor Robotics. (5 February 2017). *PaPeRo.* https://www.razorrobotics.com/robots/papero/.

[765] Osada, J., Ohnaka, S., & Sato, M. (June 2006). The scenario and design process of childcare robot, PaPeRo. In *Proceedings of the 2006 ACM SIGCHI International Conference on Advances in Computer Entertainment Technology* (pp. 80–es).

equipped to walk by itself, ensuring that safety is prioritised. PaPeRo has several technical features to cater to persons with dementia and older adults[766]:

- Interface technology for dementia with cognitive disabilities
  - Speech recognition technology
  - Information support technology
- Behaviour monitoring systems
  - Extract the decrease in cognitive disabilities
  - Activity and behaviour recognition technology
- Accurate, easy-to-understand information on older adults' day-to-day activities
- Information support system for each cognitive level
  - Familiar communication style
- Information for daily activities
  - Activities
  - Schedule reminder
  - Emergency alarm
  - Entertainment

The interactive robot can remind users of their daily routines, like when to take their medication from time to time. NEC places importance on warm and heartfelt communication through robots. Although messages or information can be delivered using a smartphone or video, PaPeRo has a communicative power not found in conventional devices. PaPeRo uses the latest speech synthesis technology, and it can speak in a natural or cute voice. It can recognise when the surroundings are noisy, or the sounds spoken to are too loud or soft. Since face-to-face communication is essential, PaPeRo, with built-in face detection, can locate people using its image recognition technology. The facial identification system can distinguish and memorise up to 30 faces. When it cannot see a face well or if the distance or location is unsatisfactory, it will self-adjust the distance or inform the user that it is too dark. The human-friendly robot can also control home electric appliances and collect online information through voice commands. It can also entertain people by dancing and playing games. PaPeRo is designed with a

[766] Research Institute of the National Rehabilitation Center for Persons with Disabilities. (n.d.). *Op. cit.*

height of 24cm, the result of careful consideration as NEC has found this is the best size for placing on tables or shelves. PaPeRo is available for US$600 monthly rent. NEC anticipated that with the recent COVID-19 outbreak, many lonely older adults can rely on PaPeRo for company.[767]

NEC has conducted demonstration experiments in several cities in Japan, including Fujieda City, Shizuoka Prefecture, Ichikawa Town, Hyogo Prefecture, and Saijo City, Ehime Prefecture.[768] Since July 2018, NEC has conducted a 'Mimamori PaPeRo' demonstration experiment in Saijo City, Ehime Prefecture. As a result, older adults sent photos to their families 1.5 times daily, and texts were exchanged thrice daily on average. The demonstration experiment reinforced the effectiveness of using robots for communication with families. In response to these positive reviews, Saijo City has begun offering the service for a fee since January 2019. NEC aims to provide "Mimamori PaPeRo" to 10,000 people by the end of 2022 through cooperation with local governments. NEC continues to work on the following deliverables[769]:

- Prototype design of robot-to-human interaction that reacts to user's needs.
- Selection of a synthetic voice that is simple to comprehend for older adults.
- Improving the robot system's voice recognition of an older adult's speech.
- Investigating the capacity of the robot system to convey information.
- Concretisation of the robot system's practical characteristics and marketing research.

---

[767] *Ibid.*

[768] NEC. (January 7 2020). *NEC、兵庫県市川町へコミュニケーション·ロボットを活用した高齢者見守りサービスを提供*. https://jpn.nec.com/press/202001/20200107_03.html.

[769] Research Institute of the National Rehabilitation Center for Persons with Disabilities. (n.d.). *An Informative and Interactive Robot to Help in Your Daily Life — Project Flow*. http://www.rehab.go.jp/ri/kaihatsu/papero_html_e/program.html.

8.4.7 *Smart Clothing that Monitors Well-being Through IoT*

Another next-generation robotics is the innovation of smart clothing that monitors well-being through IoT (Figure 8.3). Researchers in Japan have been developing innovative types of fabric fused with technology to assist the population to better cope with ageing. The development ranges from infant clothes with baby monitors to clothes for adults that allow real-time health data collection. For example, smart tags placed in shoes and socks may assist in tracking dementia patients who wander frequently, and launderable ultra-high frequency (UHF) labels in medical clinic bedlinens can notify administrators when it is time to dispose of them. According to Tokyo-based consultancy, Fuji Keizai, Japan's smart clothing market will grow to approximately US$66 million in 2022, 11 times bigger than in 2017.[771]

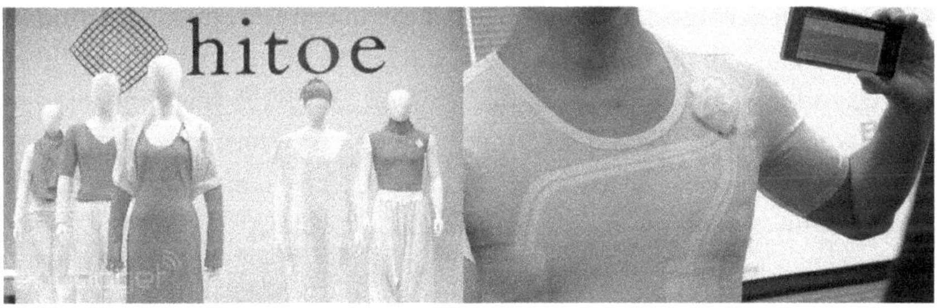

Figure 8.3. Clothing infused with Hitoe technology to track health through mobile phone.

Source: Engadget.[770]

[770] Smith, M. (30 January 2014). Japanese "Smart Clothing" Uses Nanofibers to Monitor Your Heart-rate. *Engadget*. https://www.engadget.com/2014-01-30-ntt-docomo-toray-smart-cloth.html.

[771] Iwano, M. (2 August 2020). From Gamers to Dogs, Japan Forges Ahead in Smart Clothing. *Nikkei Asia*. https://asia.nikkei.com/Business/Technology/From-gamers-to-dogs-Japan-forges-ahead-in-smart-clothing

Hitoe, which means 'one layer' in Japanese, is a functional material that can take continuous measurements of biological information.[772] NTT DoCoMo, a Japanese mobile carrier, and Toray, a material developer, have produced this material for smart sportswear and also in small hats, pyjamas, and everyday clothing. As the fabric operates as a sensor, Hitoe can collect and transmit all physiological electrical data that touch it. If made into a shirt, it can monitor an individual's heart rate; if fused into a watch, it can measure muscular data; and if worn as a hat or bandana, the technology can check one's brain activity. Hidden under a transmittable layer, the sensors are the size of a square nicotine patch, designed to be less time-consuming and less invasive than the usual electrocardiogram (ECG) test found in healthcare facilities. It can measure changes in heart rate over long periods and in different circumstances. An individual's activity can be classified as lying down, standing, sitting upright, or walking based on calculations acquired from the three-axis accelerometer in the transmitter attached to the garment.

Combined with other sensors, such as accelerometers, this product can be expanded in various ways. The health measurements are then synced to the health application platform on DoCoMo smartphones. The smart wear technology is also durable enough to withstand washing and rugged use, making it practical and functional monitoring hardware. Apart from expanding its current healthcare-related services, the two companies aim to expand into nursing and rehabilitation care. In tandem with the government's plan to reduce the number of hospital beds for acute care, individuals who wear Hitoe can cultivate the practice of rehabilitation at home as the functional material can show the patient's physical condition and provide signs if there is any change in condition when individuals are recovering at home. Offering high-quality rehabilitation through smart clothing is another innovation to tackle ageing population issues.[773]

The role of the government is critical and essential in the development and innovation of the robotic technologies industry. Dr Shibata from the

[772] Ogasawara, T., Matsunaga, K., Ito, H., & Mukaino, M. (2018). Application for rehabilitation medicine using wearable textile "hitoe". *NTT Technical Review*, *16*(9), 6–12.
[773] *Ibid.*

National Institute of Advanced Industrial Science and Technology, who developed Paro, acknowledged that robotic technologies owe much of their success to government support; Paro was created with approximately US$20 million from state funds.[774]

Japan's national R&D programme for robotics is planned and funded by the Economic Revitalisation Policy and Science, Technology, and Innovation Policy.[775] To ensure standardisation in the system, the government lays out the Robot Industry Development Plan to establish a robot innovation centre and build a national robot testing and evaluation centre. According to METI, these research centres will develop core robotic components, such as motors, drives, and high-precision reducers.[776] Currently, these components are secured through imports. Producing them locally will allow the country to lower cost as these components constitute about 60–70% of total development cost.

Investment-wise, in 2015, METI had allocated US$45 million in subsidies to urge the industry's growth.[777] In 2017, the MHLW invested another US$50 million to integrate robot technologies into 5,000 elderly care facilities. As of 2018, 36 of 47 prefectures in Japan, Tokyo, included, offer subsidies to improve healthcare and long-term care services. METI projected that its domestic market for care robots will expand to US$3.8 billion by 2035 compared to US$19.2 million in 2016. As of 2017, 26% of 857 nursing homes have reported using a robot, an increase from 2016 at 17.6%. Of the 17.6% (from 2016 data), 13% relied on transfer assistance robots like RIBA-II (robot for interactive body assistance), while the remaining used interactive robots, such as Paro and PaPeRo, that offer

[774] Foster, M. (27 March 2018). Aging Japan: Robots May Have Role in Future of Elder Care. *Reuters*. https://www.reuters.com/article/idUSKBN1H33AB/.

[775] International Federation of Robotics. (2021). *World Robotics R&D Programs*. https://ifr.org/downloads/papers/Executive_Summary_-_World_Robotics_RD_Programs_V02.pdf.

[776] Lim, T. W. (2015). *Robotic Technologies in Japan: Narratives of Use of Technologies for Manpower Enhancement, not Replacement*. Lewiston; Lampeter: The Edwin Mellen Press.

[777] Lim T. W. (2020). *Robotics in the Elderly Care Sector: A Comparative Study Between China and Japan.* EAI Background Brief No. 1531. https://research.nus.edu.sg/eai/wp-content/uploads/sites/2/2020/12/EAIBB-No.-1531-Robotics-in-elderly-care-sector-in-China-and-Japan-2.pdf.

comfort and interaction with residents.[778] The emerging evidence suggests encouraging responses from older adult users.

A survey conducted by Orix Living, a private eldercare company in Tokyo, found that over 80% of respondents expressed positive views about the use of robots in nursing care; 13.1% showed a more enthusiastic response and would like to receive nursing care from robots, while 71.2% said they are willing to receive such care if recommended.[779] Another study by Harada revealed that respondents felt comfortable communicating with robots and that the interaction was natural.[780] But, they do not feel the same towards faceless software, demonstrating the importance of robot appearance in making them feel comfortable. Yet another study, specifically on PaPeRo, informed that respondents who left verbal messages to PaPeRo, without prior interaction experience, wanted to use the robot again and thought that the interactive robot was cute.[781]

There is also a benefit to healthcare workers; the adoption of robots in nursing homes reduces stress on nurses and increases management quality among senior care workers.[782] A survey on the effectiveness of care robots found that 42% of care workers who had used them reported that the robots reduced their psychological burden, while 32% said that the number of visits to residents' rooms had reduced, allowing them to do other

[778] Eggleston, K., Lee, Y. S., & Iizuka, T. (2021). *Robots and labor in the service sector: Evidence from nursing homes* (No. w28322). National Bureau of Economic Research. https://doi.org/10.3386/w28322.

[779] Suwa, S., Tsujimura, M., Ide, H., Kodate, N., Ishimaru, M., Shimamura, A., & Yu, W. (2020). Home-care professionals' ethical perceptions of the development and use of home-care robots for older adults in Japan. *International Journal of Human–Computer Interaction, 36*(14), 1295–1303.

[780] Harada, T. E. (June 2002). Effects of agency and social contexts faced to verbal interface system. In *Proceedings of the 19th Japanese Cognitive Science Society Annual Meeting (In Japanese)* (pp. 14–15).

[781] Komatsu, T., Iwaoka, T., & Nambu, M. (2006). Leaving a message with the PaPeRo robot: The effect of interaction experience with real or virtual PaPeRo on impression evaluation. In R. Harper, M. Rauterberg, & M. Combetto (Eds.), *Entertainment Computing — ICEC 2006* (pp. 37–42). Springer. https://doi.org/10.1007/11872320_5.

[782] Eggleston *et al.* (2021). *Op. cit.*

non-strenuous tasks.[783] Although robots are being utilised in various settings and for multiple purposes, the costs of robots are high and remain not affordable for the individual; the prices of these technologies could continue to be a roadblock to their adoption in Japan.[784] Care facilities and nursing homes will have to rely on local and government subsidies to finance the cost of robots, while individuals who need it outside of facility care must claim through nursing care insurance. This raises the question of their sustainability within the general population in the long run.

Another concern put forward by Dr. Kazuhiro Kosuge, a robotics professor at Tohoku University, is that Japan's expectations for robots are too high, especially when there is still a limited number of specialist engineers who can build systems that enable robots to work in the larger society.[785] This also applies to the supply of people who can adjust, service, and repair robots. These skill shortages can potentially constrain the further development and adoption of robots and smart wear in everyday life. The implication is that the prospect of robots in addressing Japan's ageing challenge may become limited, based on hope rather than current reality.

A related issue is that current robot technologies cannot fully take on human responsibilities or match human performance. Japan's Henn na Hotel, launched in 2015 with a staff of robots, has reduced its robotic workforce by half after the experience failed to reduce costs or workload for its employees.[786] Some robots cannot answer guest questions the way digital assistants, such as Siri and Alexa can, while others require humans to step in and help. Even though it may be a riveting experience for hotel

---

[783] Obayashi, K. & Masuyama, S. (2020). Pilot and feasibility study on elderly support services using communicative robots and monitoring sensors integrated with cloud robotics. *Clinical Therapeutics*, *42*(2), 364–371.

[784] Inagaki, K. (8 October 2014). Japan aims to turn robotics into profit. *The Financial Times*. https://www.ft.com/content/e98a5d08-4ae1-11e4-839a-00144feab7de.

[785] Soble, J. (27 October 2014). The Silver Economy: Japan Embraces Future of Robot Care. *The Financial Times*. https://www.ft.com/content/cdb0dbe8-292b-11e4-8b81-00144feabdc0.

[786] Hertzfeld, E. (31 January 2019). *Japan's Henn Na Hotel Fires Half Its Robot Workforce*. Hotel Management. https://www.hotelmanagement.net/tech/japan-s-henn-na-hotel-fires-half-its-robot-workforce.

guests, it seems the robot technology is not entirely up to running and fully automated. Perhaps this is a useful reminder that the robots are not there to replace humans but to assist.

In sum, the critical success factors for next-generation robots include the following[787]:

- **Policy-driven governmental frameworks**: It is essential to develop clear and supportive governmental guidelines that address safety, privacy, and accountability and promote technological advancements. These frameworks, such as the New Robot Strategy, facilitate the integration and continuous innovation of robotics technology into the public and private sectors while ensuring they meet high safety and ethical standards.
- **Interdisciplinary and intersectoral collaboration**: Effective development and integration of next-generation robots require cooperation across various fields, including healthcare, technology, and public policy. Moreover, the partnerships between government agencies like METI and MHLW, along with the involvement of private sector companies and research institutions, are crucial for advancing robotic technologies. Specifically, subsidies, incentives, and funding from both government and private sectors can help overcome financial barriers to adoption.
- **Ethical considerations**: Addressing ethical concerns such as privacy, autonomy, and the potential dehumanization of care is essential. Promoting social acceptance through public education about the benefits and limitations of robotic assistance is critical for widespread adoption.
- **User-centric design with adaptability and scalability**: Robots are designed with the end-user in mind, ensuring they are user-friendly, accessible, and tailored to the needs of diverse populations to different

[787] Nambu, T. (2016). Legal regulations and public policies for next-generation robots in Japan. *AI & SOCIETY*, *31*(4), 483–500; Ikeda, Y., Kobayakawa, M., Nakao, H., Iseki, F., & Matsushita, H. (2021). Information technology/artificial intelligence innovations needed for better quality of life in caregiving homes. In H. Matsushita (Ed.), *Health Informatics: Translating Information into Innovation* (pp. 37–58). Springer.

tasks and environments, including older adults with limited mobility. In this light, scalability by addressing the diverse needs of users is crucial for the economic production and widespread deployment of robotic solutions.

The introduction and widespread adoption of next-generation robots also face several notable challenges[788]:

- **High costs**: The development, production, and maintenance of advanced robotic systems can be costly. This makes it difficult for many sectors, particularly smaller-scale operations or universal public services, to afford these technologies.
- **Social concerns**: Robots in caregiving roles or social interactions may raise questions about the potential for reduced human contact and deteriorated mental well-being. There is also concern about robots making errors that could have severe consequences in sensitive environments, such as healthcare.
- **Technological literacy**: The effective use of robots requires a certain level of technological literacy among the general population, healthcare workers, and caregivers. A steep learning curve may be associated with these technologies, particularly among older adults or those with limited socioeconomic resources.
- **Data security**: Robots that collect and process personal data pose significant risks to data security and privacy. Ensuring data protection against hacking and misuse is a critical challenge.

8.5 Conclusion

The development of next-generation healthcare systems in Japan is a testament to the country's commitment to integrating advanced technology into healthcare management and service delivery. Tokyo's policy-driven approach underscores the significance of a robust governmental framework in steering national healthcare agenda towards digital transformation and innovation. Specifically, the inception of this transformative journey

---

[788] *Ibid.*

can be traced to the government's E-Japan Strategy in 2001, followed by policy enhancements through the E-Japan Strategy II, New IT Reform Strategy, and I-Japan Strategy 2015. These strategic frameworks laid the groundwork for the 2018 Growth Strategy, which aims to create a healthcare system that not only extends life expectancy but also provides personalised care to promote individuals' health.

Substantial financial investment and a collaborative partnership between the public and private sectors are critical for implementing and scaling technological innovations in healthcare. Government funding and subsidies have been instrumental in promoting the adoption of advanced healthcare technologies, including robotics in facilities and digital infrastructure development for older adults. This financial support, coupled with collaborative efforts among technology developers, healthcare providers, and policymakers, is crucial for the sustained integration and optimisation of digital and robotic technologies in healthcare services.

Infrastructure development, particularly establishing 5G networks, has been a cornerstone of Japan's healthcare modernisation efforts. The deployment of 5G is instrumental in facilitating high-speed, large-capacity, and low-latency telecommunication services essential for effective telemedicine, remote surgical operations, and comprehensive health and medical information platforms. In particular, the Ministry of Internal Affairs and Communications (MIC) has been pivotal in orchestrating this technological upgrade, ensuring the infrastructure is in place to mitigate the challenges posed by the uneven distribution of medical resources. The journey towards next-generation healthcare systems and robotics is a multi-faceted endeavour, involving strategic policy frameworks, public–private partnerships, and robust infrastructure to create a more efficient, personalised, and technologically advanced healthcare system in the face of population ageing.

CHAPTER 9

VANCOUVER: SMART GOVERNANCE

9.1 City Overview

Location	Located on the western half of Burrard Peninsula, on the mainland of British Columbia
Land Size	114 km^2
Climate	Moderate, oceanic climate
Geography (topology/ landscape)	A coastal, seaport city enclosed to the north by English Bay and Burrard Inlet and to the south by Fraser River
Environmental Risks (if any)	Climate change
Population	662,248 (2021)
GDP	US$121.1 billion (Metro Vancouver region, 2020)
GDP Per Capita	US$48,087 (Metro Vancouver region, 2020)
Key Urban Issues of Smart City	Insufficient scale of start-ups, social inclusion

Metro Vancouver is the metropolitan area and most densely populated region in British Columbia, Canada. It consists of 21 municipalities and covers an area of 2,882.55 km^2.[789] Its major urban centre is the City of Vancouver (Vancouver hereafter). Located on the western half of Burrard Peninsula, Vancouver is a coastal, seaport city.[790] Vancouver is composed of 22 distinct neighbourhoods and covers an area of 114 km^2.[791] Vancouver has a moderate, oceanic climate where winters are relatively mild and rainy, and summers are sunny.[792] The mean temperature in Vancouver is 9.5°C, where the lowest temperature is in December at 2.7°C while the highest is in August at 18.2°C.[793] Vancouver receives 2,351 mm of rainfall per year, with the greatest amount of precipitation occurring in November and the lowest in July. With climate change, the city has been experiencing modest sea level rise and moderate coastal flooding events.[794]

Metro Vancouver has the third largest GDP in Canada, with a GDP of C$163.8 billion[795] (US$124.1 billion) and a GDP per capita of C$63,453 (US$48,087) in 2020. Vancouver's regional economy is characterised by service and technology.[796] Vancouver is the largest job centre in the Metro Vancouver region, with 20% of businesses located within the city,

[789] Statistics Canada. (8 February 2012). *Focus on Geography Series, 2011 Census.* https://www12.statcan.gc.ca/census-recensement/2011/as-sa/fogs-spg/Facts-cma-eng.cfm?LANG=Eng&GK=CMA&GC=933.

[790] City of Vancouver. (n.d.a). *Geography.* https://vancouver.ca/news-calendar/geo.aspx.

[791] *Ibid.*; City of Vancouver. (n.d.b). *Areas of the City.* https://vancouver.ca/news-calendar/areas-of-the-city.aspx.

[792] City of Vancouver. (n.d.c). *Weather in Vancouver.* https://vancouver.ca/news-calendar/weather.aspx; Weatherbase. (n.d.). *Vancouver, British Columbia.* http://www.weatherbase.com/weather/weather-summary.php3?s=29817&cityname=Vancouver,+British+Columbia,+Canada.

[793] Climate Data. (n.d.). *Vancouver Climate.* https://en.climate-data.org/north-america/canada/british-columbia/vancouver-963/.

[794] City of Vancouver. (n.d.d). *Climate Change and Sea Level Rise.* https://vancouver.ca/green-vancouver/climate-change-and-sea-level-rise.aspx

[795] Statista. (n.d.). *GDP at constant prices Vancouver, British Columbia.* https://www.statista.com/statistics/652758/real-gdp-of-vancouver-british-columbia/.

[796] City of Vancouver. (2020a). *Economic Structure of Vancouver: Vancouver's Role: From Provincial to International.* https://vancouver.ca/files/cov/1-4-economic-structure-vancouvers-role-from-provincial-to-international.pdf.

providing 33% of the region's jobs (2016 Census).[797] The majority of jobs are found within the information and cultural industries; professional, scientific, and technical services; and management of companies and enterprises.[798]

Vancouver has a population of 662,248 and a population density of 5,749.9 per square kilometre (2021), making it the most densely populated municipality in Canada.[799] The city is expected to grow to 856,830 by 2050.[800] Vancouver is a young community with an average age of 42.2 years old. Most of the population is in the 15–64 age group (478,960), followed by the 65 and over age group (128,415) and the 0–14 age group (70,570).[801] The population is composed of a diverse mix of ethnicities: Chinese (159,080), English (84,370), Scottish (69,200), and Irish (65,695). This diversity results from waves of immigration that have occurred over time. Over 274,360 immigrants live in Vancouver, the majority from China, Hongkong, Philippines, India, Taiwan, and Iran.

The City of Vancouver Council sets the direction and vision for the city's future.[802] The city council is led by an elected mayor and 10 council members, serving a 4-year term.[803] During this term, the city council is responsible for city services, such as bylaws, taxes, streets, sidewalks, and

---

[797] City of Vancouver. (2020b). *Economic Structure of Vancouver: Role of Vancouver in the Metro Vancouver region*. https://vancouver.ca/files/cov/1-2-economic-structure-vancouvers-role-in-the-region.pdf.

[798] City of Vancouver. (2020c). *Economic Structure of Vancouver: Employment Characteristics of Vancouver*. https://vancouver.ca/files/cov/1-1-economic-structure-employment-characteristics-of-vancouver.pdf; City of Vancouver. (n.d.e). *Hazards that Could Affect Our City*. https://vancouver.ca/home-property-development/prepare-for-other-disasters.aspx.

[799] Statistics Canada. (9 February 2022). *Census profile, 2021 Census of Population*. https://www12.statcan.gc.ca/census-recensement/2021/dp-pd/prof/index.cfm?Lang=E.

[800] Paul Mochrie. (2022). *City of Vancouver 2021 Census Population and Dwelling Counts and Metro Vancouver Growth Projection Update*. https://vancouver.ca/files/cov/2022-02-28-2021-census-population-dwelling-counts-and-metro-van-growth.pdf.

[801] Statistics Canada. (9 February 2022). *Op. cit.*

[802] City of Vancouver. (n.d.f). *Your Government*. https://vancouver.ca/your-government.aspx.

[803] City of Vancouver. (n.d.g). *Vancouver City Council*. https://vancouver.ca/your-government/vancouver-city-council.aspx.

water, while the Vancouver Board of Parks manages 250 parks and public recreation programmes within the city.[804]

9.2 Smart City Development: Smart, Connected, and Green City

In 2013, Vancouver introduced its smart city vision — to be a smart, intelligent, connected, and green city by using technology and data to improve connection and convenience.[805] The vision of a smart city was catalysed by several factors, including external pressures, new innovations, new data sources, social and environmental needs, operational constraints, and changing citizen expectations. Vancouver has established various key smart city strategies in its[806]:

- Greenest city 2020 action plan.
- Transportation 2040.
- Healthy city strategy.
- Congestion management strategy.
- Digital strategy 2022.

In 2011, the city council approved the Greenest City 2020 Action Plan.[807] The plan outlines 10 goal areas and 15 measurable targets that will help Vancouver become the greenest city in the world by 2020. Three high-level, aspirational goals — zero carbon emissions, zero waste, and healthy ecosystems — are supported by subgoals and target areas to create an integrated strategy. In June 2015, the city released a midpoint update on the goals, including plans to shift to 100% renewable energy. Progress

---

[804] City of Vancouver. (n.d.h). *Vancouver Board of Parks and Recreation*. https://vancouver.ca/your-government/vancouver-board-of-parks-and-recreation.aspx.

[805] Jessie Adcock. (2017). *Smart City the City of Vancouver Digital Journey*. https://vancouver.ca/files/cov/smart-city-vancouver-digital-journey-jessie-adcock-slideshow-dec2017-web.pdf.

[806] City of Vancouver. (2019). *Smart Cities Challenge Joint Application*. https://vancouver.ca/files/cov/2019-03-04-smart-cities-challenge-joint-application.pdf.

[807] City of Vancouver. (2015a). *Greenest City 2020 Action Plan Part Two: 2015–2020*. https://vancouver.ca/files/cov/greenest-city-2020-action-plan-2015-2020.pdf.

updates on the annual implementation were also made available on web platforms.

In 2012, the city council approved Transportation 2040, a long-term strategic plan to guide the city's future transportation and land use decisions and public investments. Transportation 2040 builds upon the Greenest City 2020 Action Plan and supports several other local, regional, and provincial policies. The plan sets specific targets to achieve a smart and efficient transportation system:

- **Zero traffic-related fatalities**: Vancouver's ultimate road safety goal is to eliminate all fatalities and serious injuries from its transportation system. Vulnerable road users, including those who utilise active transportation options and at-risk demographics including children, older adults, and individuals with mobility issues, will receive special attention.
- **Increased mode share**: The city would increase the sustainable transportation mode share to 50% of all trips by 2020. By 2040, at least 66% of trips will be made by foot, bike, and transit.
- **Reduce driving distances**: By 2020, the average distance driven per resident will decrease by 20% from 2007 levels.

In 2015, Vancouver's Healthy City Strategy, a 4-year Action Plan, was unanimously approved by the city council.[808] The plan outlines concrete measures the city will take to advance its Healthy City for All vision and establishes a comprehensive, integrated, and measurable course for the city's future. Three major areas of attention include the following:

- Healthy people — taking care of the basics.
- Healthy communities — cultivating connections.
- Healthy environments — ensuring liveability now and into the future.

The plan comprises 13 long-term goals, along with targets and indicators to monitor progress and results over a 10-year period.

---

[808] City of Vancouver. (2015b). *A Healthy City for All*. https://vancouver.ca/files/cov/Healthy-City-Strategy-Phase-2-Action-Plan-2015-2018.pdf.

The 2018 Congestion Management Strategy builds upon the Transportation 2040 plan and provides direction on identifying, monitoring, and developing strategies to address congested areas.[809] Five key priority goals with specific targets include the following:

- Improve monitoring of traffic conditions and trends.
- Improve road safety.
- Ensure a smart and efficient transportation system.
- Coordinate street use.
- Prioritise people and goods movement.

The strategy also explores the implementation of new technology to improve the transportation network (e.g., video analytics to proactively study intersections with potential conflicts, smart street lighting, and responsive signal timing systems).

Vancouver's Digital Strategy sets out the key vision for Vancouver's digital future — to enhance multi-directional digital connections among citizens, employees, businesses, and government.[810] It seeks to enhance and expand Vancouver's digital infrastructure and strengthen the city's digital economy and society to transform Vancouver into a smart city and a global leader in innovation and growth. The Digital Strategy has four main pillars:[811]

- Pillar 1: Engagement and access.
- Pillar 2: Infrastructure and assets.
- Pillar 3: Digital Economy.
- Pillar 4: Organisational digital maturity.

The city will develop, attract, and retain talent and business in the digital sector by creating enabling infrastructure, such as a favourable regulatory environment, a digital district, incubator, and cluster platform,

---

[809] City of Vancouver. (2018). *Congestion Management Strategy*. https://vancouver.ca/files/cov/congestion-management-strategy.pdf.

[810] City of Vancouver. (2013). *City of Vancouver Digital Strategy*. https://vancouver.ca/files/cov/City_of_Vancouver_Digital_Strategy.pdf.

[811] *Ibid.*

manpower training, and a mature and citizen-centric digital culture to support digital business opportunities.[812] As with many other Canadian cities, Vancouver's economic challenge is in scaling up start-ups into large, profitable companies.[813] Several roadblocks may be identified, including limited scale of the economy, the lack of sufficient domestic venture capital, and the loss of human capital.[814] A strong leadership and leading by example are considered essential to establish digital governance, for creating accountability, digital standards, and processes. A key strategy is to enhance and implement smart practices in local government. The City Council proactively develops a working culture that encourages and supports staff to innovate with digital technologies.

Vancouver Smart City Timeline

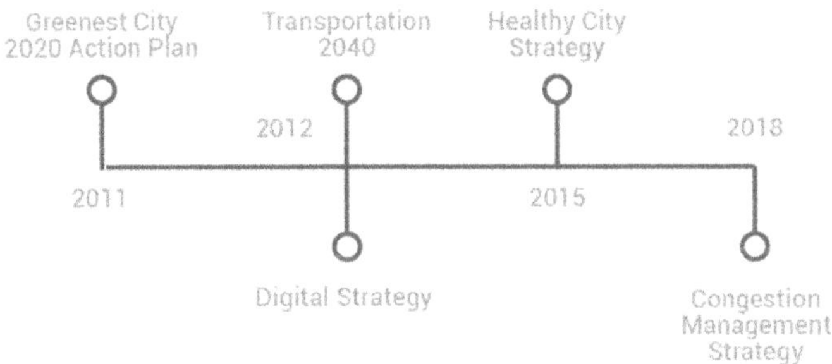

9.3 Smart Governance

Technology is fast changing how cities provide services and engage with citizens. The citizens are a dynamic group — citizen expectations are constantly shifting, and individuals increasingly desire and expect a seamless and personalised experience in their everyday life. Recognising that the provision of digital technology is a key consideration in the economic

[812] City of Vancouver. (n.d.i). *Vancouver Innovation Economy Map*. https://vancouver.ca/home-property-development/vancouver-innovation-economy-map.aspx.

[813] Hutchison, C. & Lin, L. W. (2021). *The Growth of Vancouver as an Innovation Hub: Challenges and Opportunities. 54*. https://commons.allard.ubc.ca/cgi/viewcontent.cgi?article=1709&context=fac_pubs; Statista. (n.d.). *Vancouver Global Business Cities Report 2022*. https://www.statista.com/study/66208/vancouver/.

[814] Healy, T. (23 June 2022). Interview with Chief Technology Officer City of Vancouver [Personal communication].

competitiveness of cities,[815] Vancouver has explored ways to improve citizen services, build a robust connected infrastructure, collaborate with the city's local business sector, and utilise technology to facilitate and enhance the work processes of city council staff.[816]

To spearhead the Digital Strategy, Vancouver hired its first-ever Chief Digital Officer (CDO) to lead the implementation of the Digital Strategy. Ms Jessie Adcock held the role from 2016 and 2019.[817] In 2019, she assumed overall responsibility for the city's technology portfolio, on an acting basis, becoming Canada's first Chief Technology Officer (CTO).[818]

When the city first envisioned the Digital Srategy, a 5–7-year plan was imagined. A condensed version of the approach emerged as the team considered organisational agility and the "fast pace of digital". The digital strategy was developed within a seven-month timeline and four development phases including the following[819]:

- **Phase 1**: In October 2012 (Figure 9.1), the Digital Strategy team worked with PwC Canada to assess the city's state of digital readiness based on four digital capability areas: online, mobile, social and infrastructure and data. The categories were measured across four different levels of maturity; 'absent', 'exploring', 'enabled', and 'connected'. In the assessment, Vancouver's digital maturity is positioned between 'exploring' and 'enabled'. Compared to other cities, such as Boston, London, and Seattle, it became evident that the city needed to improve

[815] Alizadeh, T. (2015). A policy analysis of digital strategies: Brisbane vs. Vancouver. *International Journal of Knowledge-Based Development*, *6*, 85–103. https://doi.org/10.1504/IJKBD.2015.071469.

[816] City Innovators Forum. (2018). *The Importance of Not Being Outpaced by Technology and the Community on the Smart City Journey.* https://cityinnovatorsforum.com/the-importance-of-not-being-outpaced-by-technology-and-the-community-on-the-smart-city-journey/.

[817] Vancouver Tech Journal. (2021). *Jessie Adcock to Join Finning as VP of Digital.* https://www.vantechjournal.com/p/jessie-adcock-finning.

[818] IT Business. (2017). *Vancouver's Chief Technology Officer Keeping the City Ahead of the Digital Curve.* https://www.itbusiness.ca/news/vancouvers-chief-technology-officer-keeping-the-city-ahead-of-the-digital-curve/95603.

[819] City of Vancouver. (n.d.j). *Timeline of Digital Strategy Achievements*. https://vancouver.ca/your-government/digital-strategy-achievements.aspx; Alizadeh (2015). *Op. cit.*

in several areas to achieve digital maturity. The city then researched and learnt from the experiences of different nations' digital initiatives. Prominent nations that have made significant progress in creating strong federal digital strategies include Canada, Australia, New Zealand, and the United Kingdom.[820]

- **Phase 2**: From November 2012 to January 2013, the team consulted industry insiders, digital thought leaders, start-up founders, business executives, citizens, and city staff. The consultation led to the formation of a 15-member external advisory committee to provide a combination of industry best practices, innovation, and academic insights to the creation of the digital strategy. The team held focus group conversations, with 45 different service providers and tech experts to delve into digital inclusion, access and e-government opportunities, infrastructure, and digital policies. Afterwards, the team conducted ideas generation with 730 individuals of the public, exploring online engagement with their municipality, followed by 35 in-depth interviews with employees across the broad range of departments in the city council. Citizens were invited to vote and comment on the top Digital Strategy ideas.
- **Phase 3**: During this phase, the city aimed to shift its digital maturity from 'exploring' to 'connected' in all four digital capability areas by the end of 2016. During January and February 2012, the digital strategy team assessed more than 120 initial ideas that emerged from phase 2 consultations and refined them into 50 concepts. From the 50 concepts, 15 initiatives were distilled and eventually categorised into 9 priority themes. The nine priorities were categorised into four pillars (Table 9.1). The first three pillars focused on enhancing the digital economy for citizens and businesses while the fourth pillar focused on the city as a digital organisation. The four pillars of Vancouver's digital strategy are as follows:
    - Engagement and access.
    - Infrastructure and assets.
    - Digital economy.
    - Organisational digital maturity.

---

[820] Healy (23 June 2022). *Op. cit.*

- **Phase 4**: In April 2013, the team presented the digital strategy to the city council. The strategy was referred to the Standing Committee on City Finance and Services, where it was officially approved by the city council. In July 2013, the city implemented the 3-1-1 Call Services for citizens to report issues. In September 2013, the city council created its first CDO position to lead the implementation of the Digital Strategy.

All nine initiatives under the four pillars were not fully funded by the city council. The three initiatives under Pillar 1 — engagement and access — were partially supported, and further funding was required. Pillar 2 — infrastructure and assets — did not have any funding and required external funding. External funding was needed for two of the initiatives under Pillar 3, digital economy. Two initiatives under Pillar 4 — organisational digital maturity — were in the process of development and required further funding. Vancouver's digital efforts were recognised in the 2018 IT World Canada Digital Transformation Conference and

Table 9.1. Four pillars and nine priority themes of Vancouver's digital strategy.

Pillar	Theme
Engagement + Access	Enable City services across digital platforms
	Expand the open data program
	Promote digital activity through communications and engagement tools
Infrastructure & Assets	Expand digital access throughout the city
Economy	Establish a digital incubation program
	Create a favourable regulatory environment that supports digital industry
	With community and industry partners, support an agile proof-of-concept program
Organizational Digital Maturity	Establish digital services governance
	Implement mobile workforce strategy

Source: City of Vancouver.[821]

[821] City of Vancouver. (n.d.k). *Timeline of Digital Strategy Achievements*. https://vancouver.ca/your-government/digital-strategy-achievements.aspx; Alizadeh (2015). *Op. cit.*

Awards, where the city won a national award for its digital strategy, under the large public sector transformation category.[822]

9.4 Implementation and Impact of Smart Governance

Vancouver has implemented many initiatives to enhance multi-directional digital connections among citizens, employees, businesses, and government. This section elaborates on the implementation and impact of two of these innovations: VanConnect 311 mobile app and VanWifi.

9.4.1 *VanConnect 311 Mobile App*

VanConnect, introduced in May 2015, is a mobile app developed by the city council to create a city service ecosystem. The app collects service requests for the city and fulfils Pillar 1 of the Digital Strategy — engagement and access.[823] The main goal of this pillar is to ensure citizens and businesses can easily interact with the city council and its services through digital channels. A secondary goal is to provide access to technology and education to increase digital literacy.[824] Prior to the establishment of the app, service requests for the city council were mainly delivered through Vancouver's 311 Call Centre, or via the council's official email. The city council decided to build a digital ecosystem when it realised that more than 40% of traffic to their website was reached through mobile devices.[825] The app was developed to fill a service gap — for individuals who prefer to access city information through digital platforms.

The app offers a hub of information, services, and access to most of the city council's programmes. Users can perform all their city-related transactions, such as reporting abandoned vehicles on city property, litter, graffiti, or broken streetlights. Once users submit their report/request, it is

[822] *Ibid.*

[823] *Ibid.*

[824] Healy (23 June 2022). *Op. cit.*

[825] Daily Hive. (2017). *City App VanConnect a Big Move Forward for Municipal Digital Strategy*. https://dailyhive.com/vancouver/city-app-vanconnect-municipal-digital-strategy.

then funnelled directly to the appropriate city department. Users can keep track of the status of their requests via an optional Facebook login or email registration. The app also displays other information, such as maps and city announcements, allowing residents to receive real-time information about their community. Accessibility is a key feature; the app is presently available in 14 languages with the ability to add more as the demographics of Vancouver change.[826] Residents can connect and engage with the city on the go as they watch council meetings, get updates on news, and receive emergency updates via their mobile phones.[827]

At the same time, the app collects geolocated service requests that include location, type of services from a predefined list of service categories, user-generated answers to issue-specific questions, and optional unstructured feedback. VanConnect's ability to collect geographic information while tailoring data output, such as maps and announcements, demonstrates its multiple functions. App development is not a one-off activity. In late 2022, the city council released a new and improved app, Van311, to connect with city services, replacing the previous app, VanConnect[828] (Figure 9.1). The improved features include the following:

- 50+ new service request types.
- Improved service status and resolution messaging.
- Ability to search for the closest community centre, park, road closures, and other information.
- Submitting service requests without creating an account.
- Inclusion of comment forms for leaving feedback to the city council.
- Access to the app's most popular requests.

As existing VanConnect accounts are not transferable, users would have to sign up for a new Van311 account. Those without the Van311

[826] Daily Hive (2017). *Op. cit.*

[827] EIT Digital. (2018). *Digital Transformation in Greece.* https://www.eitdigital.eu/fileadmin/newsroom/publications/Digital_Transformation_in_Greece_2018.pdf.

[828] City of Vancouver. (2022a). *Report Issues and Request Services with Van311.* https://vancouver.ca/van311.aspx

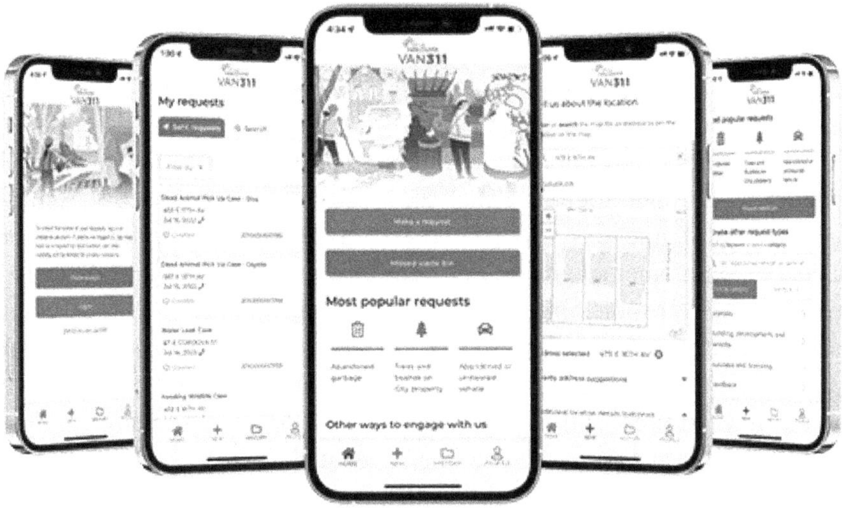

Figure 9.1. Van311 App.
Source: City of Vancouver.[829]

mobile application can report their issue or request a service on their desktop.[830] During the changeover, all service requests prior to the creation of the new Van311 app will remain available to city staff and will be worked on until completion.

The development of the Van311 app encapsulates how volunteered geographic information can help advance dialogue and consultation between citizens and government. As the Director of Digital & Contact Centre Services from the city council explained, the main aim of the app is to enhance citizen–government communication and increase the government's efficiency while creating an improved citizen experience.[831] It is noted that in Vancouver, the app does not replace traditional communication channels. While the city council was hopeful that the mobile app could replace the existing 311 call service and its corresponding high operating costs, the call and email service remained available. The

[829] City of Vancouver (2022a). *Op. cit.*
[830] City of Vancouver. (31 March 2023). *Van311*. https://van311.ca.
[831] Sangiambut, S. & Sieber, R. (2016). The V in VGI: Citizens or civic data sources. *Urban Planning*, *1*(2), 141–154.

rationale is that maintaining the traditional call and email service would help bridge digital divides among different sociodemographic groups and demonstrate the city council's commitment to ensure that everyone can receive basic services in an inclusive and accessible manner.

The Van311 app observed an increase in usage with the COVID-19 pandemic. In order not to overload the service of 911 (intended for emergencies only), the city council created a section within the app for COVID-19 enforcement requests.[832] The city also categorised COVID-19-related 311 call topics into 26 different classifications,[833] such as complaints about physical distancing violations, referrals to different organisations, and inquiries relating to the pandemic.[834] The significance of the app lies in its ability to reach a wider audience, including younger citizens. Younger individuals are often alienated and estranged from civic participation; their engagement through apps like Van311 would inform the production of public goods and services of consequence to them and facilitate participative democracy and neighbourhood empowerment.[835]

9.4.2 *Van WiFi*

The free public Wi-Fi, known as VanWifi, aims to enhance citizens' daily experience through online, mobile, and social media channels.[836] This fulfils Pillar 2 of the Digital Strategy — infrastructure and assets, which

---

[832] Lalonde, M. (22 August 2020). *City of Vancouver Rolls Out Online System to Catch COVID-19 Rule-Breakers — Vancouver is Awesome.* https://www.vancouverisawesome.com/local-news/vancouver-online-system-covid-19-rule-breakers-2657670.

[833] City of Vancouver. (n.d.l). *Open Data Portal.* https://opendata.vancouver.ca/explore/?disjunctive.features&disjunctive.theme&disjunctive.keyword&disjunctive.data-owner&++++++++disjunctive.data-team&sort=modified&q=311&disjunctive.data-team.

[834] Edith Coates & Myla White. (2020). Noisy, Nosey and Nurturing: An Analysis of Vancouver's Non-emergency Line During COVID-19. *The Ubyssey.* https://ubyssey.ca/science/311-data-covid-19/.

[835] United Nations. (2018). *Young People's Civic and Political Engagement and Global Citizenship.* United Nations. https://www.un.org/en/chronicle/article/young-peoples-civic-and-political-engagement-and-global-citizenship.

[836] City of Vancouver. (n.d.m). *Featured Project: #VanWiFi.* https://vancouver.ca/your-government/vanwifi.aspx.

aims to provide the city with robust digital infrastructure. A key strategy is to collaborate with industry.

In 2014, the city council sent out collaboration requests to all city-based telecommunications providers. Vancouver's request for external funding amplifies city efforts and harnesses external support and resources to drive its digital transformation. In April 2015, the city council selected and approved TELUS's plans to provide free Wi-Fi at 43 public facilities, including community centres, parks, outdoor pools, theatres, and cultural centres.[837] In September 2015, the first VanWiFi location was completed for a beta launch at the Mount Pleasant Community Centre.[838]

The next phase of digital infrastructure development started in 2016 when Shaw Communications Inc. was selected to add public Wi-Fi to another 15 locations, including all nine city-owned social housing sites and six outdoor sites.[839] Shaw Communications exceeded expectations and offered free Wi-Fi to an additional 500 locations throughout the downtown core and surrounding areas.[840] In 2017, it was reported that free public Wi-Fi had 280,000 users.[841]

The third phase started in late 2019 and network expansion was delivered by TELUS. The aim is to fully integrate VanWiFi in plazas and improve indoor cellular coverage at the Moberly Arts and Cultural Centre.[842] This third network expansion was completed in 2020. As of 2023, there are 521 free public Wi-Fi locations across Vancouver (Figure 9.2), making VanWiFi one of the largest free public Wi-Fi networks

---

[837] Vancouversun. (2015). *Telus Gets the Go-ahead to Roll Out Free Wi-Fi in Vancouver. Vancouversun.* https://vancouversun.com/news/staff-blogs/telus-gets-the-go-ahead-to-roll-out-free-wi-fi-in-vancouver.

[838] City of Vancouver. (n.d.n). *Timeline of Digital Strategy Achievements.* https://vancouver.ca/your-government/digital-strategy-achievements.aspx.

[839] *InfoTel News.* (19 January 2018). Vancouver Hoping to Drive Tourism Dollars with Free Public Wi-Fi Expansion. *INFOnews.* https://infotel.ca/newsitem/vancouver-shaw-wifi/cp1623061738.

[840] Broadband Technology Report. (2018). *Vancouver Taps Shaw for Free Public Wi-Fi.* https://www.broadbandtechreport.com/wireless/wifi/article/16449243/vancouver-taps-shaw-for-free-public-wifi.

[841] Jessie Adcock (2017). *Op. cit.*

[842] The Link Paper (2019). *TELUS and City of Vancouver to Expand Free Public Wi-Fi.* https://thelinkpaper.ca/telus-and-city-of-vancouver-to-expand-free-public-wifi/.

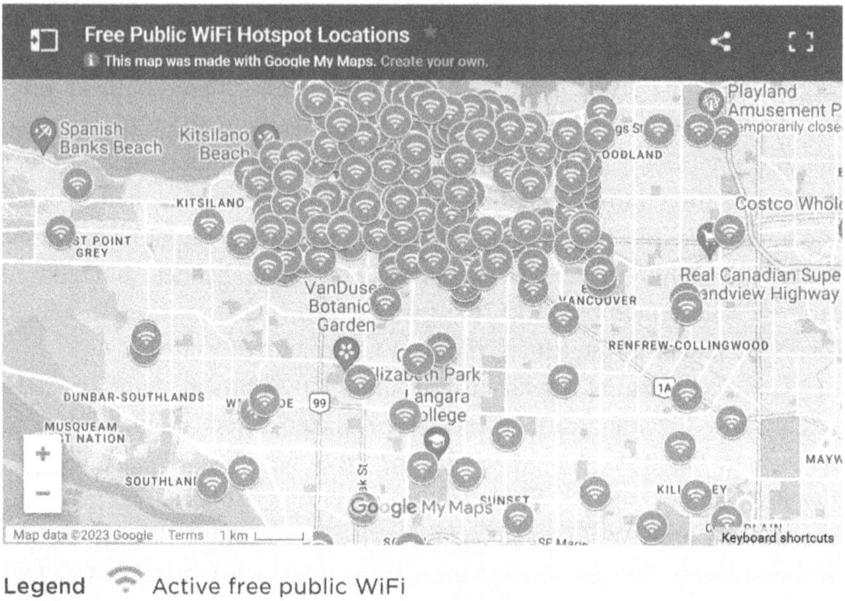

Figure 9.2. Vancouver map of locations that have free Wi-Fi.
Source: City of Vancouver.[843]

in North America.[844] Its widespread availability and easy accessibility enhance daily living.[845]

VanWiFi provides both locals and visitors with enhanced connectivity, easier communication, and convenience when out of home. The free public WiFi is easy to connect to; no personal information is required to use the network.[846] To connect, one needs only select the VanWiFi network name on the device setting menu. The browser will automatically

[843] City of Vancouver. (n.d.p). *Featured Project: #VanWiFi*. https://vancouver.ca/your-government/vanwifi.aspx.

[844] City of Vancouver. (n.d.o). *Featured Project: #VanWiFi*. https://vancouver.ca/your-government/vanwifi.aspx.

[845] Healy (23 June 2022). *Op. cit.*

[846] Inside Vancouver. (2018). Vancouver's Expanded Free Public Wi-Fi Network Among the Largest in North America. *Inside Vancouver*. https://www.insidevancouver.ca/2018/04/04/vancouvers-expanded-free-public-wifi-network-among-the-largest-in-north-america/.

redirect users to the terms and conditions page. Upon accepting the terms and conditions, users will be re-directed to the VanWiFi home page, which then connects them to the internet. Users can expect a bandwidth speed of 10 Mbps with no cap on data usage.[847]

During the COVID-19 pandemic, public health orders and social distancing restrictions were implemented. The closure of many public and commercial spaces with VanWiFi led to numerous groups being excluded from the connectivity; without internet access, many people lived without social connections and entertainment.[848] Barriers to public Wi-Fi accessibility deterred individuals from getting timely and accurate information on the pandemic and government initiatives. Open letters were sent to the city council.[849] Responding to the need for expanded access to free Wi-Fi, in 2020, Shaw Communications offered its free "Go WiFi" programme to all citizens, including non-Shaw customers.[850] Other internet providers, such as Rogers, TELUS, Bell, and TekSavvy, also suspended data limits for home internet customers.[851]

VanWiFi's successful deployment demonstrates the importance of involving different stakeholders from various city departments and organisations as well as public–private partnerships in the development of the city's digital infrastructure and digital economy.[852] At present, there are no official data on the effectiveness and usability of VanWiFi among the local community. However, it is noted that the availability and accessibility of the Wi-Fi locations enable visitors to use their smartphones and devices to

---

[847] Broadband Technology Report (2018). *Op. cit.*

[848] City of Vancouver. (2022b). *Vancouver Budget 2022*. https://vancouver.ca/files/cov/2022-budget-service-plans.PDF.

[849] Kwan, J. (2020). *Need for Expansion of Free Wi-Fi Access in the Downtown Eastside*. https://www.jennykwanndp.ca/need_for_expansion_of_free_wifi_access_in_the_downtown_eastside.

[850] Bench, A. (15 March 2020). *Shaw Opens Up 'Go Wi-Fi' Hotspots to Canadians for Free During COVID-19 Outbreak*. https://globalnews.ca/news/6680500/shaw-covid-19-wifi-opens-up-free/.

[851] Breen, K. (13 March 2020). *Coronavirus: Internet Providers Suspend Data Overage Fees for Home Internet*. https://globalnews.ca/news/6673413/coronavirus-rogers-data-overage-fees/.

[852] Healy (23 June 2022). *Op. cit.*

discover attractions, activities, and restaurants, for wayfinding, and to navigate Vancouver's transportation options.[853]

The implementation of smart governance through the VanConnect Mobile App and VanWifi has its challenges including the following[854]:

- **Digital literacy and education**: There is a need to address digital literacy and provide digital education within the city as many residents may not be familiar with the use of mobile apps or accessing Wi-Fi services. If not addressed, this could hinder the adoption and effectiveness of these smart initiatives.
- **Complexity of services**: Vancouver's services are often complex and layered due to years of development. Simplifying these services to make them more accessible to the public is a major ongoing challenge. Streamlining complex services is essential for effective implementation of smart governance solutions, such as VanConnect.
- **Procurement and development**: Public–private partnerships are helpful to smart governance solution development but there are challenges, especially when dealing with external across-sector tasks such as early-stage development or influencing the direction of product development.
- **Logistics and legal agreements**: Implementing smart initiatives such as VanWifi involves logistical challenges, such as determining access point locations and negotiating legal agreements with service providers, such as Shaw and TELUS.
- **User experience and feedback loop**: Ensuring a smooth user experience and establishing an effective feedback loop between service requests and end-users is critical. However, there are challenges in maintaining a clean and smooth feedback loop, which requires continuous improvement efforts.
- **Strategic planning and growth**: As these smart initiatives mature, there is a need for strategic planning to assess their growth potential and the additional value they can bring to the city. Revisiting strategic

[853] *InfoTel News*. (19 January 2018). *Op. cit.*
[854] Healy (23 June 2022). *Op. cit.*

plans periodically is essential to ensure that these initiatives continue to meet evolving citizen needs and expectations.

Despite the challenges, there are critical factors that support the success of these initiatives[855]:

- **Partner ecosystem and collaboration**: Scaling initiatives like VanConnect were achieved through a partner ecosystem approach involving internal and external partners. Leveraging pre-approved vendors (private sector) for user experience design and testing allowed for flexible scaling up and down as needed. Collaboration extends across city departments, e.g., including engineering and various business units, for a whole-of-government approach.
- **Talent pool and academic partnerships**: Vancouver's access to a talented workforce from local universities and design institutions has contributed to the success of smart governance initiatives. Academic partnerships helped foster talent production, attracting multinational tech companies, and building up industries such as VR and AR within the city.
- **Strategy focused on tangible outcomes**: The strategy for smart governance initiatives was anchored in delivering tangible outcomes, such as digital service delivery and improved access to city services for citizens. This focus on delivering benefits to citizens helped drive adoption and support.
- **Insulation from operational complexities**: Insulating the team behind initiatives like VanConnect from the complexities of IT back-office operations allowed for a sharper focus on advancing strategic initiatives. This separation enabled the team to dedicate more resources and attention to driving innovation and improving user experience.
- **Iterative product development**: Continuous feedback and iteration based on user behaviours and preferences were key to the success of VanConnect. Analysing user feedback and usage patterns helped in building a product roadmap focused on enhancing the app's primary

---

[855] *Ibid.*

function, such as service request submission, to improve overall user satisfaction and reputation.

9.5 Conclusion

Vancouver's Digital Strategy showcases a multi-directional focus involving a wide domain, including e-governance, e-infrastructure, e-business, and e-democracy.[856] The strategy started off by adopting an international benchmark that compares the city's digital readiness against digitally advanced cities in the world. The 4-year strategy, implemented in 2013, has enabled Vancouver to go from having a low digital maturity to being a model for cities around the world. The development of a connected city has led to strong citizen support and a more integrated and strategic governance approach that brings value to citizens and businesses.[857] The journey has not ended as the city continues to take its digital footprint to the next level.

Vancouver's approach to smart governance is people-focused. Strong leadership and collaboration, operationalised through the city government and its public services, ensures that the city has sufficient capacity and resources to plan and execute city-wide digital transformation strategies. The development of the Van311 app demonstrates an application of volunteered geographic information in fostering dialogue between citizens and government. By enhancing communication and efficiency, the app improves the responsiveness of governance as well as citizen experiences. Even so, the city council continues to maintain both digital and traditional channels of service communication to cater to differing levels of technological comfort and literacy, ensuring inclusivity and accessibility to all.

Initiatives like VanWiFi further democratise access to the city's digital infrastructure, fostering connectivity and engagement among citizens, including the younger demographic, which is often alienated from civic participation. The pandemic has accelerated Vancouver's smart city initiatives, highlighting the importance of digital technologies in providing critical support and addressing citizens' needs during crises. The Van311

[856] Alizadeh (2015). *Op. cit.*

[857] Healy (23 June 2022). *Op. cit.*

app has a designated section for COVID-19-related requests, helping to alleviate pressure on emergency services. By prioritising digital strategies, the city not only adapts and responds to pandemic challenges but also builds transparency and trust between citizens and government. Through public–private partnerships and stakeholder involvement, Vancouver's digital infrastructure expansion enhances both connectivity and the city's commitment to social inclusion and economic growth.

CHAPTER 10

COPENHAGEN: CLIMATE RESILIENCE ACTION

10.1 City Overview

Location[858]	Copenhagen — Capital City of Denmark
Land Size[859]	90.1 km^2
Climate[860]	Temperate maritime
Geography (topology/ landscape)[861]	Coastal location, flat terrain, extensive waterways, and green spaces
Environmental Risks (if any)[862]	Water pollution (due to industrial discharge and solid waste disposal)

[858] City of Copenhagen. (n.d.a). *An Internationally-connected Region.* https://international.kk.dk/about-copenhagen/an-internationally-connected-region.

[859] StatBank Denmark. (n.d.a). *Area 1. January.* https://www.statbank.dk/statbank5a/Graphics/mapanalyser.asp?maintable=ARE207&lang=1.

[860] Intergovernmental Panel on Climate Changes. (2014). *Information for Participants.* https://www.ipcc.ch/site/assets/uploads/2018/05/Copenhagen_Info_0714.pdf.

[861] City of Copenhagen. (n.d.b). *Københavns Kommune: Indre By.* https://web.archive.org/web/20131023035646/http://www.kk.dk/da/brugbyen/brug-steder-i-byen/groenne-omraader/parker/indre-by.

[862] Madsen, H. M., Mikkelsen, P. S., & Blok, A. (2019). Framing professional climate risk knowledge: Extreme weather events as drivers of adaptation innovation in Copenhagen, Denmark. *Environmental Science & Policy*, *98*, 30–38.

Population[863]	656,787 (2023)
GDP[864]	DKK 517.115 billion (US$74.45 billion) (2021)
GDP Per Capita[865]	DKK 645,000 (US$92,856) (2021)
Key Urban Issues of Smart City	Climate change, energy source

Copenhagen is Denmark's capital and most populous city. It is situated on the eastern coastal plain of Zealand, the largest island in Denmark, characterised by low-lying fields, forests, and wetlands. Copenhagen's low-lying topography and proximity to the sea make it susceptible to flooding, coastal erosion, storm surges, and sea-level rise.[866] The Copenhagen Municipality has a land area of 90.1 km^2 [867] and a population of 656,787[868] as of 2023. With urbanisation trends, Copenhagen's metropolitan area is witnessing escalated population growth, accentuating the need for smart solutions to manage urban complexities and preserve liveability. Copenhagen experiences a temperate maritime climate, with mild winters, cool summers (the average temperature of the warmest months is 20°C), and a fair amount of precipitation throughout the year.

Copenhagen is the cultural, economic, and financial centre of Denmark. In 2021, Copenhagen had a GDP per capita of DKK 645,000,[869] employing 414,363 people.[870] The city's economy is based mainly on services and commerce. Other notable industries include life science[871] and

[863] StatBank Denmark. (n.d.b). *Population at the First day of the Quarter.* https://www.statistikbanken.dk/statbank5a/Graphics/mapanalyser.asp?maintable=FOLK1A&lang=1.

[864] Statista. (n.d.a). *Copenhagen: GDP 2021.* https://www.statista.com/statistics/1304689/copenhagen-gdp/.

[865] Statista. (n.d.b). *Copenhagen: GDP per capita 2021.* https://www.statista.com/statistics/1304716/copenhagen-gdp-per-capita/.

[866] Intergovernmental Panel on Climate Changes. (2014). *Op. cit.*

[867] StatBank Denmark. (n.d.b). *Op. cit.*

[868] *Ibid.*

[869] Statista. (n.d.b). *Op. cit.*

[870] StatBank Denmark. (n.d.c). *Average Number of Employed.* https://www.statbank.dk/statbank5a/Graphics/mapanalyser.asp?maintable=KAS301&lang=1.

[871] Copenhagen Convention Bureau. (6 March 2023). Copenhagen: Life Science Congress Capital for Global Health. *POLITICO.* https://www.politico.eu/sponsored-content/copenhagen-life-science-congress-capital-for-global-health/.

maritime sector.[872] Smart city innovation is pursued to bolster the city's economy by attracting tech companies, start-ups, and knowledge institutions to drive investments and create jobs.

10.2 Smart City Development: A Carbon-Neutral and Resilient City

Copenhagen's smart city vision is to develop "an attractive 'smart city' with a strong environmental profile" by creative and experimental solutions.[873] This vision is embedded into three main objectives[874]:

- Achieving carbon neutrality to become the world's first carbon-neutral capital by 2025.
- Creating a greener, more sustainable, and more liveable capital city.
- Supporting economic growth.

To fulfil the objectives, Copenhagen has developed a series of smart city initiatives including the following:

- **Transition to renewable energy**: Copenhagen aims to primarily rely on renewable energy for power and heat generation by 2025,[875] in line with the Danish energy policy mandates by 2050.[876] Renewable energy currently accounts for 30% of all energy used in Denmark,[877] encompassing wind, biomass, geothermal energy, and waste. The transition

[872] Ministry of Foreign Affairs. (n.d.). *The Maritime Industry in Denmark — A Leading Hub in Europe*. https://investindk.com/set-up-a-business/maritime.

[873] City of Copenhagen. (n.d.c). *Ambitious Growth Policies. https://international.kk.dk/about-copenhagen/ambitious-growth-policies.*

[874] HEC Pairs. (n.d.). *Smart Cities — The Sustainable Program of Six Leading Cities*. https://www.hec.edu/en/smart-cities-sustainable-program-six-leading-cities.

[875] City of Copenhagen. (2012). *The CPH 2025 Climate Plan*. https://kk.sites.itera.dk/apps/kk_pub2/index.asp?mode=detalje&id=983.

[876] Global Climate Action Partnership. (n.d.). *Transformational Change: Danish 100% Renewable Energy Policy*. https://globalclimateactionpartnership.org/case-study/transformational-change-danish-100-renewable-energy-policy-2/.

[877] Denmark.Dk. (n.d.). *Pioneers in Clean Energy*. https://denmark.dk/innovation-and-design/clean-energy.

to renewable energy is not just a mere infrastructural shift but also an opportunity for Copenhagen towards sustainable development, ensuring health, quality of life, mobility, and green growth in the metropolis.[878]

- **Enhancing climate resilience**: As a coastal city, Copenhagen is increasingly exposed to climate risks, such as sea-level rise, frequent flooding, and extreme weather events.[879] Copenhagen seeks to build climate resilience through a range of measures, such as the transition to renewable energy mentioned above, promoting sustainable transportation options, and developing climate-resilient neighbourhoods. In particular, the Climate Resilient Neighbourhood initiative focuses on adopting green solutions in different neighbourhoods in Copenhagen, including developing green spaces, creating urban water management systems, and implementing climate-adaptive building design.[880]
- **Green and smart mobility**: Copenhagen has launched a series of initiatives to strengthen green and smart mobility, improving urban quality of life and growth potential. The city has been actively pursuing the development of the City of Cyclists, aiming to enhance bicycle connections and promote the use of bicycles (traditional and electric).[881] An innovative platform, Copenhagen Intelligent Traffic Solutions (CITS), is developed to improve traffic flow, reduce emissions, and enhance safety for citizens by utilising a network of Wi-Fi access points to geolocate Wi-Fi-enabled devices. The data gathered through the system enables city officials to monitor real-time traffic conditions and simulate the design of transport-related tools and apps to improve urban mobility.[882]

---

[878] City of Copenhagen. (2012). *Op. cit.*

[879] Negrello, M. (2023). Designing with nature climate-resilient cities: A lesson from Copenhagen. In E. Arbizzani, E. Cangelli, C. Clemente, F. Cumo, F. Giofrè, A. M. Giovenale, M. Palme, & S. Paris (Eds.), *Technological Imagination in the Green and Digital Transition* (pp. 853–862). Springer International Publishing.

[880] *Klimakvarter Østerbro.* (n.d.). *The Climate-resilient Neighbourhood Østerbro.* https://klimakvarter.dk/en/.

[881] City of Copenhagen. (2012). *Op. cit.*

[882] Copenhagen Intelligent Traffic Solutions (CITS). (n.d.). *State of Green.* https://stateofgreen.com/en/solutions/cits-copenhagen-intelligent-traffic-solutions/.

- **Green economy**: Copenhagen won the European Green Capital Award in 2014 and is considered a model for the green economy.[883] The local government collaborates with research institutions and start-up communities to create the Copenhagen Science City in the heart of the city, viewed as one of the strongest innovation districts in Europe.[884] The Nordhavn district is another example of smart neighbourhood with a full-scale smart city energy lab demonstrating how electricity, heating, buildings and transportation can be integrated into an intelligent, flexible energy system.[885]
- **Cross-sectoral partnerships**: Collaborations between the public sector, private companies, and academic institutions further promote Copenhagen's smart urban development. For instance, the Copenhagen Solutions Lab was founded in 2014 to support the city's smart city development through testing and implementing intelligent and data-driven solutions.[886] It acts as a bridge between external partners and the City of Copenhagen to develop and implement smart city initiatives, such as the Smart Cities Accelerator+ and building energy optimisation using artificial intelligence.[887]
- **Digital transformation of municipal services**: Copenhagen integrates advanced digital technologies to improve infrastructure and municipal services, including using a data-driven approach for more effective urban planning and resource management. In 2016, Copenhagen's City Data Exchange was launched by the Hitachi Insight Group, marking the world's first big data platform for cities.[888]

[883] European Commission. (n.d.). *About the Awards.* https://environment.ec.europa.eu/topics/urban-environment/european-green-capital-award/about-awards_en.

[884] City of Copenhagen. (n.d.d). *Copenhagen Science City. https://international.kk.dk/about-copenhagen/copenhagen-science-city.*

[885] State of Green. (n.d.). *Nordhavn: The Smart Urban Area of the Future.* https://stateofgreen.com/en/news/nordhavn-take-a-tour-of-an-urban-area-of-the-future/.

[886] Copenhagen Solutions Lab. (n.d.a). *People and Flows.* https://cphsolutionslab.dk/en.

[887] Copenhagen Solutions Lab. (n.d.b). *Urban Laboratories.* https://cphsolutionslab.dk/en/projekter/samarbejder.

[888] State of Green. (n.d.). *City of Copenhagen Launches the World's First Big Data Platform for Cities.* https://stateofgreen.com/en/news/city-of-copenhagen-launches-the-worlds-first-big-data-platform-for-cities/.

The platform hosts private and public data to facilitate the creation of enhanced public solutions and new business opportunities.[889]

Copenhagen Smart City Timeline

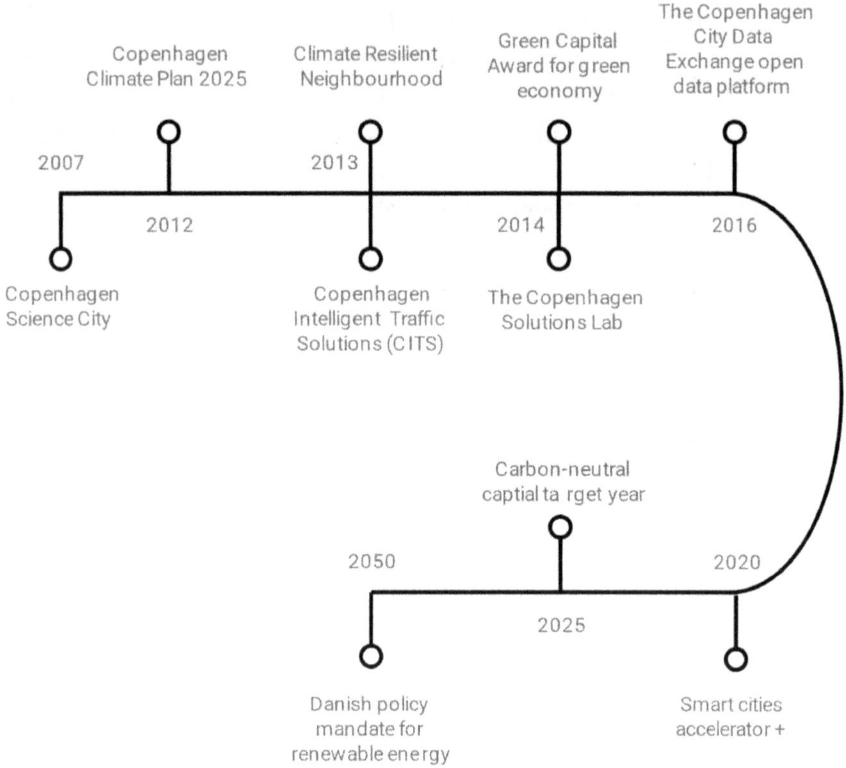

10.3 Climate Resilience Action

Climate resilience is a critical aspect of urban sustainability, particularly in coastal cities like Copenhagen, which are vulnerable to sea level rise and extreme weather events. On 23 August 2012, the CPH 2025 Climate Plan was proposed by the City of Copenhagen, aiming to achieve carbon neutrality by 2025 by addressing four pillars[890]:

---

[889] *Ibid.*

[890] City of Copenhagen (2012). *Op. cit.*

- Energy consumption.
- Energy production.
- Mobility.
- City administration.

In addition to transferring relevant initiatives from the 2009 Climate Plan, the CPH 2025 Climate Plan contains objectives, primary efforts, and initiatives adopted by the city council to achieve the goal of carbon neutrality. The vision is for Copenhagen to become a global leader in urban sustainability, demonstrating how big cities can create green growth and development while reducing carbon emissions simultaneously. The plan also seeks to enhance Copenhageners' quality of life by using the carbon neutrality goal as a lever to change behaviours and stimulate innovation, job creation, and investments.[891]

Collaboration with strategic partners, such as government agencies, universities, and companies, is a key strategy for achieving the carbon neutrality goal. Copenhagen encourages Danish and international operators to develop new solutions and technologies in the city. For example, Energy Leap is a partnership involving the City of Copenhagen, HOFOR, and over 40 property owners. It focuses on optimising energy usage in buildings, including electricity management for ventilation and large-scale catering.[892] ATELIER is a five-year EU Horizon 2020 project to convert buildings from energy consumers to producers. It encourages collaboration between authorities, residents, businesses, and knowledge institutions to develop urban infrastructure for energy-positive urban areas. Collaborators include Copenhagen, Bilbao, and Amsterdam. A collaboration forum facilitates early resource sharing among the cities.[893]

The CPH 2025 plan would require Copenhagen to raise approximately DKK2.7 billion (US$400 million) leading up to 2025 to implement its initiatives.[894] The investment is expected to lead to an overall

---

[891] *Ibid.*

[892] State of Green. (2018). *Energy Data Ensure Better Buildings*. https://stateofgreen.com/en/solutions/energy-data-ensure-better-buildings/

[893] ATELIER. (n.d.). *Objectives*. https://smartcity-atelier.eu/about/objectives/

[894] Nikel, D. (22 August 2019). Copenhagen Aims to Be World's First Carbon Neutral Capital by 2025. *Forbes*. https://www.forbes.com/sites/davidnikel/2019/08/22/copenhagen-

positive economic impact and direct benefits for the residents, e.g., improved health and quality of life for Copenhageners. More than half of the expected cost (approximately DKK1.4 billion) would be used to retrofit city buildings to improve energy efficiency. Another DKK270 million has been earmarked for replacing street lighting, DKK600 million for investment in cycling infrastructure, and DKK290 million for public transport.[895] The 2025 Climate Plan will be implemented in three phases:

- Roadmap 2013–2016.
- Roadmap 2017–2020.
- Roadmap 2021–2025.

Phased development allows the Climate Plan to be evaluated at the end of each phase leading up to 2025 and to further develop with the emergence of new knowledge and technology, potentially utilising the most updated lessons learnt as well as adapting to the continuously changing socioeconomic conditions.

10.3.1 *Climate Resilient Neighbourhoods*

In July 2011, Copenhagen experienced its worst cloudburst in 400 years, with 152.4 mm of rain in less than three hours.[896] The entire deluge caused DKK 6 billion (US$1.04 billion) in damages.[897] In 2014, another cloudburst of similar intensity hit Copenhagen and Malmø regions, affecting the northern outskirts and missing the city centre.[898] The Climate-Resilient

---

aims-to-be-worlds-first-carbon-neutral-capital-by-2025/?sh=7c6a23bf7fb3.

[895] City of Copenhagen. (2012). *Op. cit.*

[896] *Denmark Keeps on Dithering Over Climate Adaptation Plans.* (9 March 2023). *Clean Energy Wire.* https://www.cleanenergywire.org/news/denmark-keeps-dithering-over-climate-adaptation-plans.

[897] Gerdes, J. (31 October 2012). What Copenhagen Can Teach Cities About Adapting to Climate Change. *Forbes.* https://www.forbes.com/sites/justingerdes/2012/10/31/what-copenhagen-can-teach-cities-about-adapting-to-climate-change/.

[898] Urban Sustainability Exchange. (n.d.). *Copenhagen Climate-resilient Neighbourhood Strategy.* https://use.metropolis.org/case-studies/copenhagen-climate-resilient-neighbourhood-strategy.

Neighbourhood initiative is part of Copenhagen's overall strategy for district and urban renewal in 2010 through the Copenhagen Climate Adaptation Plan, seeking to safeguard Copenhagen and its population against the risks of future extreme weather events.[899]

St. Kjelds is a diverse and densely populated neighbourhood with few green areas. The district houses 24,000 residents, and over 10,000 jobs are concentrated in 1 square kilometre.[900] St. Kjelds was selected as a pilot neighbourhood where methods and technologies for climate adaptation would be developed. Collaborators include the City of Copenhagen's Climate Department, HOFOR, planners, architects, landscape architects (e.g., Tredje Natur, SLA Architects, and Malmos Landskaber), and local community.[901] The integrated urban renewal project in St. Kjelds was allocated a total budget of DKK60 million (US$ 8.7 million), from the former Ministry of Housing and the City of Copenhagen. Of this DKK60 million, DKK48.45 million (US$ 6.9 million) was allocated to the St. Kjeld's Square and Bryggervangen redevelopment.[902] Specifically, green solutions at the street level have been implemented in St. Kjelds to lead stormwater to the harbour, where it can be dispersed without causing any damage to the city landscape. The goals include the following:

- To prevent flooding in the neighbourhood of St. Kjelds by sustainably managing the stormwater/rainwater through blue solutions on the surface.
- To create more green spaces and enhance biodiversity.
- To develop a stronger neighbourhood identity by bringing tangible solutions to the citizens.
- To present a scalable climate change strategy that can inspire other cities.

[899] City of Copenhagen. (n.d.e). *Climate Adaptation.* https://international.kk.dk/about-copenhagen/liveable-green-city/climate-adaptation.

[900] State of Green. (n.d.). *Copenhagen's First Climate Resilient Neighbourhood.* https://stateofgreen.com/en/solutions/copenhagens-first-climate-resilient-neighbourhood/.

[901] *Ibid.*

[902] Urban Sustainability Exchange (n.d.). *Op. cit.*

10.4 Implementation and Impact of Climate Resilience Action

This section elaborates on the implementation and impact of two climate resilience initiatives in Copenhagen: Copenhagen's CPH 2025 Climate Plan and Climate Resilient Neighbourhood.

10.4.1 *The CPH 2025 Climate Plan*

The City of Copenhagen adopts a holistic approach to implementing the CPH 2025 Climate Plan towards achieving carbon neutrality by 2025. Specifically, Copenhagen has instituted energy conservation measures, such as building code revisions and retrofitting, to streamline the city's energy consumption, aligning with best practices in sustainable urban planning. The city's pivot towards renewable energy sources, including wind and biomass, exemplifies a systematic and sustainable energy transition. The CPH 2050 plan identifies several strategic directions to guide the implementation including the following[903]:

- **Reducing heat consumption**: Copenhagen's heat consumption has stabilised since 2010. This unchanged total heat consumption results from the growth (2.6 million m2) of the built-up area of Copenhagen, equating to roughly 5% of the building stock in 2019. Positively, heat consumption has not increased due to the initiatives implemented by the Greater Copenhagen Utility (HOFOR) since 2010 to optimise district heating units and streamline the district heating systems. For example, HOFOR's efforts in converting the steam-powered segment of the district heating grid to water-based heat have noticeably reduced heat loss. Also, efforts were made in urban development areas to reduce the flow temperature of district-heating supply. These initiatives, combined with day-to-day efforts to optimise the temperature in the combined district heating grid, have decreased the need for heat production and reduced carbon emissions.

[903] City of Copenhagen (2012). *Op. cit.*

- **Reducing electricity consumption**: Copenhagen's electricity consumption started to decline in 2010 but has remained unchanged and even slightly increased in more recent years. Energy retrofitting was a crucial Roadmap 2017–2020 initiative to reduce electricity consumption. However, it was abandoned as the City of Copenhagen could only support energy retrofitting in buildings owned by the city, 5% of Copenhagen's building stock. For the rest (95%) of the building stock, Copenhagen can only support refurbishment in the social housing sector (10% of Copenhagen's building stock) and provide urban renewal funds to privately owned buildings. Furthermore, many new buildings being built in Copenhagen frequently use more energy in practice than stipulated in the energy requirements of the building code. In response, Copenhagen is seeking to optimise energy in building stock through the following:
    - Statutory requirements mandating energy labelling of all buildings built before 2000 and upgrading their energy label to a specific minimum over several years.
    - Ensuring better compliance with the building code.
    - Enhancing financial incentives to save energy.
    - Reducing electricity at Copenhagen's utility companies. BIOFOS, the wastewater utility, is expected to save electricity in wastewater treatment processes, while HOFOR is expanding district cooling, which is more energy-efficient than individual cooling systems, thus reducing overall electricity consumption.
- **Reducing household energy consumption**: The City of Copenhagen has worked to remove barriers and establish conditions for investments in energy retrofitting while considering architecture, indoor climate, and enhanced daylight. For example, in 2019, the City of Copenhagen issued grants to support 16 properties containing 907 housing units, with an expected average energy savings of 23% per property. Support in 2018–2020 particularly focused on replacing windows with energy-efficient, noise-insulating windows. Support was granted to more than 2,000 housing units in this same period. The climate plan promoted green retrofitting of social housing in the same period. In May 2020, the Danish government agreed to earmark

DKK30.2 billion for green retrofitting of housing units in the social housing sector up to 2026. The agreement aims to ensure healthy, contemporary social housing for the benefit of tenants, green transformation, and the restoration of the Danish economy. The City of Copenhagen expects that 42 projects can be initiated with a total capital expenditure of DKK2.5 billion between 2020 and 2021. The project will comprise 11,000 housing units in Copenhagen and provide scope for significant improvements (with a green focus) in social housing.

- **District heating and sorting out plastic waste**: In 2019, HOFOR's new biomass-fired power station unit, BIO4, started operation at Amager Power Station, supplying district heating to Copenhagen. BIO4 replaces Copenhagen's last coal-fired power station unit, and all heat and power production in the city is now converted from coal to biomass. This has resulted in 80% of Copenhagen's district heating being carbon neutral. The remaining 20% of the heat is generated from fossil fractions in waste and by the city's peak load and reserve load plants — most of which are fired with oil and natural gas. The main challenge is non-biodegradable waste, such as plastics, which can produce carbon emissions when incinerated along with biomass for district heating or energy production. In response, the city council adopted the Circular Copenhagen: Resource and Waste Management Plan 2024 in 2019, stipulating several initiatives supporting the CPH 2025 Climate Plan's goal to remove plastic and biodegradable waste from the rest of household waste. Copenhagen and Amager Resource Centre have actively removed plastic in collaboration with citizens and Copenhagen companies. For the same reason, it became possible in 2017 for Copenhageners to sort hard and soft plastic for recycling. Currently, 3,700 tonnes of plastic waste are sorted for recycling in Copenhagen each year, equating to 23% of the total volume of plastic waste. The target is to sort 13,000 tonnes of plastic waste for recycling by 2025, equating to 86% of the total volume of plastic. The Resource and Waste Management Plan also included an initiative to establish a biogas plant close to the city for biodegradable waste. Altogether, the initiatives contribute to increasing production of biogas capable of replacing the use of fossil (natural) gas and reducing the volume of plastic in waste incineration, thereby reducing carbon emissions in Copenhagen's energy systems.

- **Wind power**: As of 2019, over 47% of total electricity consumption in Denmark was by wind power. In 2019, 16.1 terawatt hours of wind power were produced in Denmark.[904] Copenhagen's current target is to have 460 MW of wind turbine capacity by 2025. To achieve this, HOFOR continues to establish onshore wind turbines in Copenhagen and outside the city boundary. In addition, HOFOR and the City of Copenhagen are establishing large offshore wind farms, preferably as close to the city as possible. Feasibility studies were conducted in 2019 and 2020, indicating two areas in Øresund (Nordre Flint and Aflandshage) which can potentially contain 410 MW of wind turbine capacity.
- **Carbon capture**: In Budget 2020, the city council noted that Amager Resource Centre is working on establishing a pilot facility that could pave the way for a full-scale carbon capture facility. Amager Resource Centre and the City of Copenhagen expect that establishing a large carbon capture facility could render waste incineration carbon neutral by capturing the CO_2 generated by the fossil fraction of the waste.
- **Mobility**: Copenhagen has recognised transportation as a critical sector for reducing CO_2 emissions. The main challenge in mobility involves the rising number of cars and the high number of kilometres driven by cars in Copenhagen, resulting in high carbon emissions. The city seeks to address several factors affecting Copenhagen's scope for reducing emissions from road traffic including the following:
    - Population growth in Copenhagen has increased the demand for transportation, putting pressure on the city's infrastructure. To alleviate this, greener transport options like cycle superhighways have been implemented, creating over 750 km of regional routes for cyclists. This has resulted in a 23% increase in cyclists, with 14% switching from cars. The city has improved cycle parking at stations and traffic hubs in collaboration with public transport companies.

[904] Statista. (n.d.c). *Denmark: Number of Active Wind Power Turbine 2010–2021*. https://www.statista.com/statistics/990723/number-of-active-wind-power-turbines-in-denmark/.

- Limited funding, funds from parking income, currently offset in the city's block grant, could have been used to promote green mobility.
- Space limitations in urban areas, especially in the inner city and "bridge"[905] neighbourhoods, pose challenges in implementing necessary infrastructure for green mobility, such as charging stations, bus lanes, cycling paths, and bicycle parking. To reduce carbon emissions from vehicles, Copenhagen aims to replace most taxis with electric vehicles by 2025. Over 100 taxis have become emission-free by the end of 2019. The city has also committed to having emission-free buses, with some already converted to electricity. In 2025, all buses in Copenhagen are expected to be converted from diesel to zero emission.[906]

- **Training and information**: This is crucial to resident acceptance and involvement in the city's green transformation. The city has initiated action requiring broad collaboration with stakeholders in Copenhagen, including youth study programmes, educational institutions, business community, and many other parties. Examples include:
  - In autumn 2020, a Sustainable City showroom[907] was unveiled to demonstrate sustainable urban development by HOFOR and the City of Copenhagen. The target audience includes children, young people, and students as well as professionals, such as urban planners, politicians, and experts. It serves as an educational platform for learning about the initiatives to build a sustainable Greater Copenhagen.
  - Climate Ambassador Training Programme, jointly run by the Children and Youth Administration and the Technical and Environmental Administration, is offered to 7th-grade students in Copenhagen. The programme engages students as part of the citizen involvement process, equipping them with qualifications and

[905] "Bridge" neighbourhoods often refer to urban areas that connect more affluent districts with those less affluent, serving as a transitional zone regarding demographics, economic status, and urban development.

[906] State of Green. (2023). *The Zero Emission Public Transport of the Future*. https://stateofgreen.com/en/solutions/the-zero-emission-public-transport-of-the-future/.

[907] ENERGI & VAND. (n.d.). *Showroom*. https://energiogvand.dk/en/frontpage/showroom/.

technical skills aligned with lower secondary school learning targets as well as knowledgeable advice from a teenager's perspective.[908] Since 2009, 500 7th-grade students have participated in the programme.

The CPH 2025 Climate Plan is not just a climate-related strategy but embodies a holistic approach to transformative urban development, encompassing economic, social, and governance issues. The plan advocates a collaborative approach to sustainability, involving various stakeholders, including government, businesses, citizens, and NGOs, in the development and implementation process. Aligning with global carbon reduction targets, Copenhagen's implementation portends a significant decrement in urban carbon emissions, setting an example for metropolitan environmental stewardship. The cost of green investments, though substantial initially, forecasts long-term dividends through energy savings and green sector employment expansion. Also, increasing green spaces, cleaner water bodies, and a stronger focus on sustainability can lead to a better quality of life for residents.

But the road to carbon neutrality is not without its pitfalls. On 1 August 2022, the Amager Resource Centre announced that it was not eligible for national carbon capture funding, leading to Copenhagen giving up on its net-zero pledge.[909] Backtracking on Copenhagen's commitments has accelerated critical discussions on net zero emissions, casting doubts for other pledges worldwide, especially the financial and technological aspects.[910] This setback unavoidably has diminished Copenhagen's standing on the global stage and affected its ability to influence other cities and countries, including potentially weakening the global collective action on

---

[908] *Ibid.*

[909] ARC. (2022). *ARC søger ikke den statslige CO2-fangstpulje.* https://a-r-c.dk/arc-soeger-ikke-den-statslige-co2-fangstpulje/.

[910] Christiansen, K. L. & Hougaard, I. (16 September 2022). Copenhagen's Failure to Meet 2025 Net Zero Target Casts Doubt on Other Pledges. *Climate Home News.* https://www.climatechangenews.com/2022/09/16/copenhagens-failure-to-meet-2025-net-zero-target-casts-doubt-on-other-city-pledges/; Szumski, C. (23 August 2022). Copenhagen's Dream of Being Carbon Neutral by 2025 Goes Up in Smoke. *Euractiv.* https://www.euractiv.com/section/energy-environment/news/copenhagens-dream-of-being-carbon-neutral-by-2025-go-up-in-smoke/.

climate change and eroding public trust in government initiatives and leadership.[911] Several scholars have raised questions about the effectiveness of Copenhagen's carbon neutrality strategies, prompting reconsideration on how to align urban planning with degrowth principles[912]:

- A narrow definition of emissions by only counting local emissions ignores emissions associated with products and services consumed in the city.
- The emphasis on making activities more efficient instead of addressing the overall impact may just slow down the growth of environmental impact.
- Sustainability measures are often framed as ways to boost economic growth and competitiveness, which can lead to increased consumption and negative environmental impacts.

Other innovation-specific challenges include the following[913]:

- Scaling up innovations to avoid death by pilots.
- Maintaining good communication with politicians and leaders for city-wide implementation.
- Resolving privacy, cybersecurity, and data leaks.
- Harmonising the vast difference in maturity in different innovations.

Notwithstanding, there are enabling factors to support the implementation of CPH 2025 Climate Plan, e.g.,[914]

- Stakeholder engagement and partnerships to support green transitions by utilising new technologies and sharing innovations.

---

[911] Sario, F. D. (10 November 2022). Cities' Zero Emissions (over) Ambition Faces Reality Check. *POLITICO*. https://www.politico.eu/article/eu-commission-targets-cities-zero-emissions-2030-ambition-reality-check/; Langston, R. (3 October 2022). What Happens When a City Abandons Its Net-Zero Target? *Net Zero Investor.* https://www.netzeroinvestor.net/news-and-views/what-happens-when-a-city-abandons-its-net-zero-target.

[912] Krähmer, K. (2021). Are green cities sustainable? A degrowth critique of sustainable urban development in Copenhagen. *European Planning Studies*, *29*(7), 1272–1289.

[913] *Ibid.*

[914] Gate 21. (20 September 2022). Interview with Gate 21 [Personal communication].

- Facilitating an environment that easily tests new technologies.
- Pilot projects approach.
- Demand-driven innovation methods to identify citizen demands for innovations and involve universities and companies to level up innovation power.
- Lots of networks and communication to disseminate innovation and solutions.
- Clear focus, only using new technologies and IoT that can contribute to reducing CO_2 emissions.

10.4.2 *Climate Resilient Neighbourhood*

The Climate Resilient Neighbourhood initiative adopted a neighbourhood-based approach to climate resilience, recognising that local communities are often the first affected by climate change impacts.[915] Engaging with the local community to identify climate change risks and develop strategies to enhance resilience is an essential task in climate action.[916] Specifically, the initiative emphasises involvement and co-creation with local stakeholders in the planning and implementation to build trust, create ownership, and ensure the effectiveness and sustainability of resilience measures.[917]

10.4.3 *St. Kjeld's Square and Bryggervangen*

Since 2012, the residents and the business community in the area have been involved through citizens' meetings, workshops, and project groups. In 2015, SLA Architects, a local Danish architectural firm, won the project competition for the redevelopment of St. Kjeld's Square and Bryggervangen, to be completed in 2018.[918] The St. Kjeld's Square and Bryggervangen

---

[915] City of Copenhagen. (2016). *Copenhagen's First Climate Resilient Neighbourhood (2nd edition).* http://klimakvarter.dk/wp-content/uploads/2015/08/Copenhagens-first-climate-resilient-neighbourhood_WEB_low.pdf.

[916] Negrello (2023). *Op. cit.*

[917] *Ibid.*

[918] Klimakvarter Østerbro. (n.d.). *Skt. Kjelds Plads & Bryggervangen.* https://klimakvarter.dk/en/projekt/skt-kjelds-plads/.

project would demonstrate how climate-adapted urban spaces could be created to address rising rainfall. The vision was to transform St. Kjeld's Square into the green heart of the Climate-Resilient Neighbourhood (Figure 10.1), with Bryggervangen, an existing 2,300-foot long road, complementing as a cloudburst road lined with rain gardens to receive rainwater (Figure 10.2) from sidewalks and adjacent residential buildings, as well as allow excess rainwater to run off into large infiltration ponds at the square (Figure 10.3).[919] Bryggervangen would also feature a First Flush system, which uses an in-ground diverter to filter dirt and sediments from the rainwater before depositing them into the infiltration ponds.[920]

The rejuvenated space would bring nature into the neighbourhood by adopting green and blue strategies, promoting healthier lifestyles and more social interactions for residents of the Climate-Resilient Neighbourhood.[921] Starting in 2012, St. Kjeld's conducted an annual "city life" survey to evaluate how residents used public spaces in the neighbourhood. The findings showed the following[922]:

- Copenhagen's citizens were more aware of climate change and the need to adapt the city to cope with heavy rains and cloudbursts.
- Residents gained inspiration through different initiatives developed by the integrated urban renewal project and were motivated to start their own projects, contributing to the local community.
- St. Kjled's had developed an identity as the Copenhagen Climate Resilient Neighbourhood and had improved its reputation as a place to live and work.

The redevelopment went on to win the prestigious Årets Arne award in 2020.[923]

---

[919] Holmes, D. (2020). Sankt Kjelds Square and Bryggervangen wins Arne of the Year Award. *World Landscape Architecture*. https://worldlandscapearchitect.com/sankt-kjelds-square-and-bryggervangen-wins-arne-of-the-year-award/.

[920] Grigo, K. (2020). Middle of the road. *Landscape Architecture Magazine September 2020* (p. 30). http://bt.royle.com/publication/?i=670045&p=32&view=issueViewer&pp=1.

[921] Klimakvarter Østerbro (n.d.). *Op. cit.*

[922] Urban Sustainability Exchange (n.d.). *Op. cit.*

[923] Nielsen, C. F. (2020). *Ambitiøst klimaprojekt vinder Årets Arne.* https://arkitektforeningen.dk/nyheder/ambitioest-klimaprojekt-vinder-aarets-arne/.

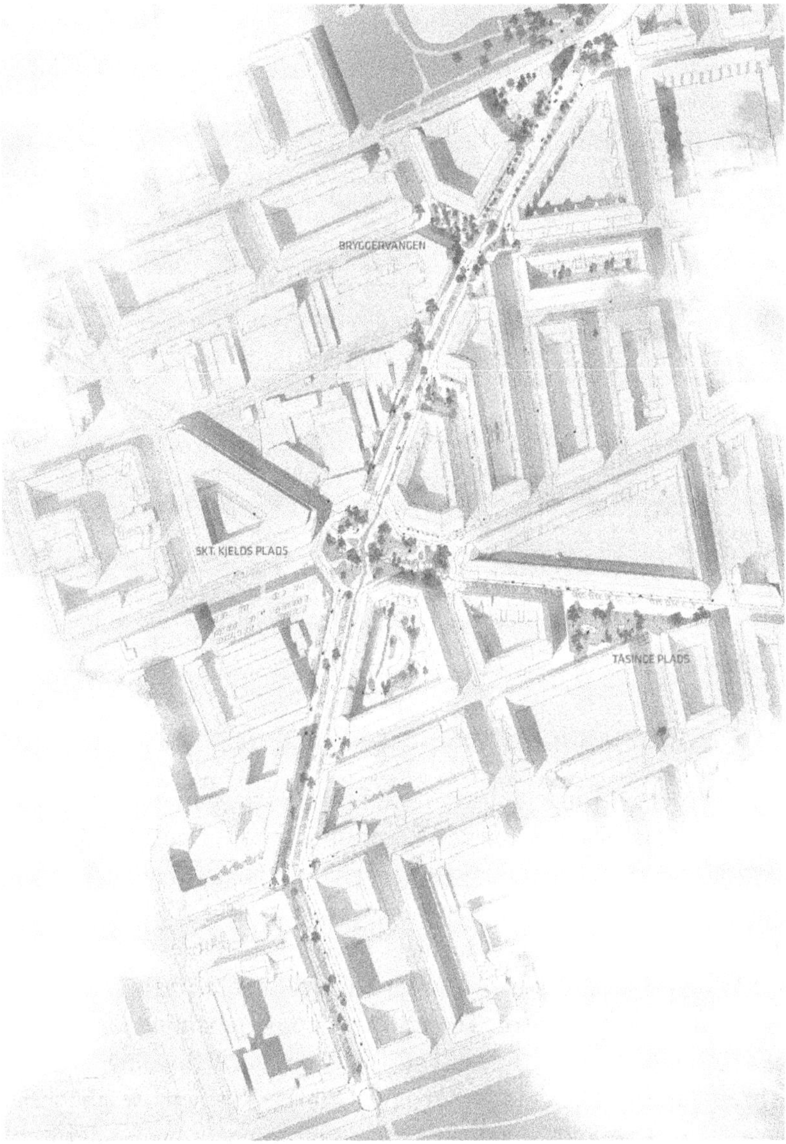

Figure 10.1. Plan drawing of St. Kjeld's Square and Bryggervangen.
Source: TREDJE NATUR.[924]

[924] TREDJE NATUR. (n.d.). *The First Climate District*. https://www.tredjenatur.dk/en/portfolio/the-first-climate-district/.

Figure 10.2. St Kjeld's Square.
Source: Holmes.[925]

10.4.4 *Tåsinge Square*

The area around Tåsinge Square in the St. Kjeld's Quarter has a history of regular flooding. The basements of surrounding bare-brickwork buildings, typical of the 1930s, are often flooded due to the accumulation of water draining off the square into the subsoil beneath the buildings. Another contributing factor is the replacement of original cobblestone path covering the subsoil with a layer of more impermeable asphalt so the square could be used as a car park. The result is a tiny remaining land area that could absorb rainwater efficiently, leading to regular flooding.[926]

In addressing this issue as part of its Climate Resilient Neighbourhood strategy, the city council initiated a design competition for projects to

[925] Holmes, D. (12 February 2020). Sankt Kjelds Square and Bryggervangen Wins Arne of the Year Award. *World Landscape Architecture*. https://worldlandscapearchitect.com/sankt-kjelds-square-and-bryggervangen-wins-arne-of-the-year-award/.
[926] PublicSpace. (n.d.). *Works*. https://www.publicspace.org/works/-/project/j075-refurbishment-of-tasinge-square.

Figure 10.3. Raingardens at Bryggervangen.
Source: Holmes.[927]

convert Tåsinge Square into a green zone, emphasising rainwater management and replacing impermeable asphalted areas with gardens, which have a greater capacity for absorbing and retaining rainwater during heavy downpours. A sum of 2 million euros (DKK14.9 million) by the City of Copenhagen and HOFOR was allocated to this renovation.[928] The competition was won by GHB Landskabsarkitekter and subsequently realised between 2013 and 2014.

The design brief provided by GHB Landskabsarkitekter seeks to develop a green oasis with nature, history, and water management-cum-play installation in the square to manage rainwater and offer learning and demonstration value. As a green oasis in a densely populated city district, the proposed design considers extreme precipitation events and equips the square with the capacity to handle heavy cloud bursts in the future through a system of retention basin and cloudburst street.[929]

[927] Holmes (2020). *Op. cit.*
[928] *Ibid.*
[929] *Ibid.*

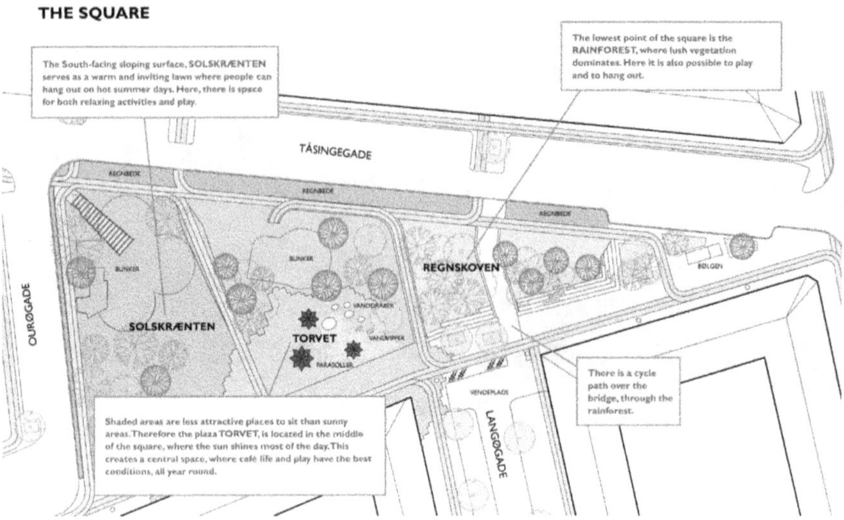

Figure 10.4. Tåsinge Square plan.
Source: Klimakvarter Østerbro.[930]

In the renovation, most of the rainwater that falls on Tåsinge Square is separated from the sewer system. Water from roofs is diverted to tanks under the Raindrops installation in the central plaza from where it will run to the rainforest and slowly infiltrate. The renovation also includes the adaptive reuse of existing infrastructure. Two air raid shelters (bunkers) have been conserved beneath the square, serving as rehearsal rooms for musicians. The entrances to the bunkers have been renovated and now serve as small seating niches, while a new staircase serves as a pathway to the grassy slope of Solskrænten (sun slope) (Figure 10.4). Tåsinge Square is designed to give the St. Kjeld's Quarter a distinctive green and sustainable landmark,[931] while serving as a new meeting place for residents.

Tåsinge Square showcases a cross-section of Danish countryside, with plant biotopes extending from the hillside to lakeside. Its 1,000-m^2 urban wilderness can delay flooding and provide drainage of rainwater

[930] Klimakvarter Østerbro. (n.d.). *Op. cit.*
[931] LYTT. (n.d.). *LYTT Architecture as.* https://www.lytt.dk/.

from the surrounding area of approximately 8,000 m^2.[932] Urban nature is implemented throughout the square, expressed through several green initiatives[933]:

- Food for birds and insects and great seasonal diversity are emphasised in the selection of plants, offering different experiences, encouraging a range of biodiversity, and creating a wilder urban nature overall.
- Plant selections consider soil conditions, sunlight, and shadow to give the vegetation the best possible growth conditions, e.g., on the square and along the streets, English oaks or flowering wild cherries have been planted in a bed of geraniums supplemented with spring flowering bulbs. As salt is used on the roads during winter to prevent ice from forming, a salt-tolerant mix of herbs and grasses is chosen for planting along the streets.

The elevated area, Solskrænten, is mostly dry, with a predominance of trees and grasses, while the lower area, the Regnskoven, features dense, lush vegetation (Figure 10.5).

The plaza in the centre of the square contains two installations representing the presence of water on the square: Rain parasols and water drops (Figure 10.6). The rain parasols collect the water and provide shelter from

Figure 10.5. Tåsinge Square vegetation.

Source: Klimakvarter Østerbro.[934]

[932] Danish Architecture Center (DAC). (n.d.). *Tåsinge Plads: A Pioneering Project in Denmark's First Climate Neighborhood.* https://dac.dk/en/knowledgebase/architecture/taasinge-plads-a-pioneering-project-in-denmarks-first-climate-neighborhood/.

[933] *Ibid.*

[934] Klimakvarter Østerbro. (n.d.). *Op. cit.*

Figure 10.6. Rain parasol and water drop installations.
Source: Authors.

the rain. The water drops installation reflects the sky and their surroundings with its bright metallic surface, inviting people to touch and climb them. Rainwater from surrounding roof surface is collected and diverted to tanks beneath the big water drops. By using two manual pumps designed as playground equipment, water is pumped out of the most significant Water drop, where it will flow on the surface and end in the rain garden (retention basin, Figure 10.7). The water running from the plaza to the rain garden meets obstacles on its way, enabling people to follow the water flow and on the way, learn and play.[935]

The Climate Resilient Neighbourhood initiative aims to achieve multiple benefits, such as[936]:

[935] *Ibid.*
[936] C40 Cities. (n.d.). *Cities100: Copenhagen — Creating a Climate Resilient Neighborhood.* https://www.c40.org/case-studies/cities100-copenhagen-creating-a-climate-resilient-neighborhood/.

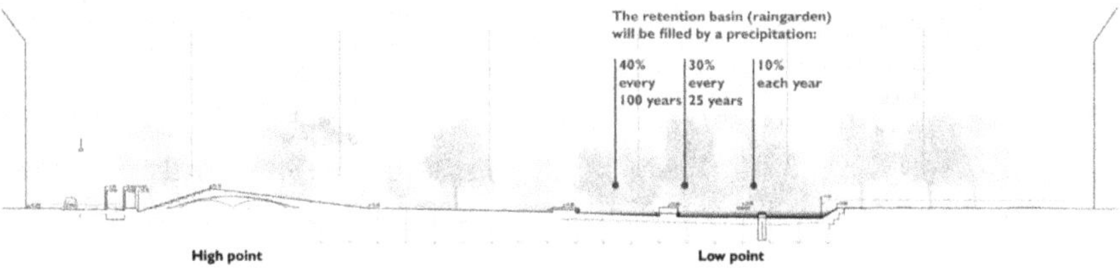

Figure 10.7. Tåsinge Square rainwater retention.
Source: Klimakvarter Østerbro.[937]

- Improving quality of life.
- Enhancing social cohesion.
- Promoting biodiversity.
- Increasing resilience to climate change.
- Acknowledging that resilience-building efforts could contribute to sustainable development by delivering multiple benefits beyond climate resilience.

Even before the new Tåsinge Square was inaugurated, residents in the neighbourhood had formed a committee to oversee the garden's maintenance and organise local activities.[938] The refurbished square is a collection of gardens with nature that offer residents a near-home meeting place where children can play and learn about rainfall. The resulting impact of replicating small green spaces all over the city for climate change adaptation could be significant for providing health, well-being, and ecological benefits[939]:

- The planted soil replacing previous asphalt surface reduces the land area once allocated for cars and increases the green cover, lowering greenhouse gas emissions.

[937] Klimakvarter Østerbro. (n.d.). *Op. cit.*

[938] Engberg, L. A. (2018). Climate adaptation and citizens' participation in Denmark: Experiences from Copenhagen. In S. Hughes, E. K. Chu, & S. G. Mason (Eds.), *Climate Change in Cities: Innovations in Multi-Level Governance* (pp. 139–161). Springer International Publishing.

[939] Liu, L. & Jensen, M. B. (2017). Climate resilience strategies of Beijing and Copenhagen and their links to sustainability. *Water Policy*, *19*(6), 997–1013.

- The growth of plants offers a land sink of carbon, improving health and air quality.
- The planted ground mitigates urban heat island effects and rainfall runoff (e.g., retention basin), reducing flooding and costly drainage networks.[940]

The strength of the initiative is in applying a holistic approach to climate resilience building through addressing extreme precipitation, flood protection, urban heat reduction, citizen engagement, and affordable green infrastructure across multiple dimensions, highlighting that resilience is not just about preventing individual hazards but also about building both place and individual adaptive capacity.[941] Investing in climate resilience building can help identify vulnerable areas and populations, ensuring that they become resilient in extreme weather events while creating co-benefits for public health, biodiversity, economic development and long-term sustainability.[942] There are pitfalls in implementation, e.g., when using an infrastructure-centric resilience strategy, which tends to exacerbate vulnerability through outcomes like eco-gentrification, necessitating inclusive, equitable development strategies, and richer community perspectives including from existing residents to foster more comprehensive resilience narratives.[943]

In sum, critical success factors for the Climate Resilient Neighbourhood initiative include the following[944]:

- **Community engagement**: For instance, the local city renewal office engaged with the public, inviting citizens to help design the renovation projects and shape its vision. Engaging with citizens promotes continuous improvement of the city, encouraging more extensive use

---

[940] PublicSpace. (n.d.). *Op. cit.*

[941] Engberg (2018). *Op. cit.*

[942] Ateş, M. & Erinsel Önder, D. (2021). A local smart city approach in the context of smart environment and urban resilience. *International Journal of Disaster Resilience in the Built Environment*, *14*(3), 266–284.

[943] *Ibid.*

[944] City of Copenhagen. (24 September 2022). Interview with the City of Copenhagen [Personal communication].

and ownership of the space. It is one of the most important tools in building smart and sustainable cities.

- **Integration of nature and urban planning**: The use of nature-based solutions, such as green spaces in flood management, provides additional benefits, including enhancing liveability, reducing air pollution, and increasing people's willingness to experience the city.
- **Consideration of biodiversity and accessibility**: Improving biodiversity and accessibility of the city invites people to use the refreshed public space and promotes environmental awareness, which are critical components of smart urban renewal.
- **Thinking boldly**: Having an ambitious goal and developing neighbourhood space informed by evidence and past mistakes improve thinking and development outcomes.

There are also several challenges including the following[945]:

- **Little and inadequate evaluation**: Evaluation is critical to understanding the perceptions and impacts from the perspective of the citizens, improving future planning and projects.
- **Funding support**: Municipal funding for building smart cities is important, requiring clear articulation of target beneficiaries and benefits, tangible and intangible to gain support from city leadership.
- **Different expectations and needs from subpopulations**: Improving communication with all and diverse citizens, including older adults, youths, and minority groups, ensures better understanding of their needs, experiences, and aspirations in project design and implementation.

10.5 Conclusion

Copenhagen's CPH 2025 Climate Plan and Climate Resilient Neighbourhood initiative both focus on the importance of a holistic approach, collaborative mechanisms, and ambitious targets for climate resilience building that effectively adapts to extreme weather events and

---

[945] *Ibid.*

improves the quality of life for all citizens. A holistic approach to climate resilience building is vital, highlighting that climate resilience is about building broader adaptive capacity (place and person), not just resolving specific climate hazards. Specifically, Copenhagen's CPH 2025 Climate Plan illustrates a comprehensive strategy, aiming for carbon neutrality by 2025, while simultaneously fostering green economic growth and enhancing the well-being of its citizens. The plan encompasses energy conservation measures, renewable energy adoption, transportation sector reforms, and green transformation engagement, showcasing multi-faceted environmental stewardship.

Stakeholder engagement is another critical learning point in both initiatives. The Climate Resilient Neighbourhood emphasises neighbourhood-based resilience, underscoring the importance of collaborative efforts involving government agencies, private sector entities, knowledge institutions, and citizens in identifying risks and developing adaptive strategies. This co-creation process fosters ownership, trust, and sustainability of resilience-building efforts. Climate resilience is a long-term endeavour and multi-actor process that requires coordination, resources, and commitment from governments, communities, businesses, and individuals, going beyond reducing carbon emissions.

Copenhagen's experience demonstrates the complexities and challenges of ambitious climate goals. The city's withdrawal from its net-zero emissions pledge due to ineligibility for national carbon capture funding underscores the financial and technological hurdles cities face in pursuing carbon neutrality. This setback not only affected Copenhagen's global standing but also illustrated the vulnerabilities and complexities inherent in implementing ambitious climate initiatives. It raises critical discussions about the feasibility and reliability of net-zero pledges, highlighting the need for realistic planning, robust support mechanisms, and transparent communication in climate action commitments.

CHAPTER 11

MELBOURNE: SMART CLIMATE ACTION

11.1 City Overview

Location	Capital city of Victoria State, Australia
Land Size	9,990 km^2
Climate	Temperate
Geography (topology/ landscape)	At the head of Port Philip Bay on the south-eastern fringe of Australian landmass
Environmental Risks	Extreme weather events from climate change impacts, e.g., heatwave, drought, and flooding
Population	4,917,750 (2021)
GDP	US$256.60 billion (2019)
GDP Per Capita	US$50,531 (2019)
Key Urban Issues of Smart City	Climate change adaptation, population growth, urban heat island effect

Melbourne is the capital city of Victoria State, Australia. The metropolitan area of Melbourne spans approximately 9,990 km^2 and consists of 31 local government areas.[946] Melbourne's climate is temperate, with warm

[946] State Government of Victoria. (2023). *Metropolitan Melbourne*. https://liveinmelbourne.vic.gov.au/discover/melbourne-victoria/metropolitan-melbourne.

summers (occasionally rising above 40°C) and cool winters (no snow). It is well known for having "four seasons in one day".[947] In the 2021 Census, Melbourne metropolitan area reached a population of 4,917,750,[948] remaining as the second most populated city in Australia after Sydney (5,231,147).[949]

In 2019, Melbourne's GDP reached A$369 billion (US$256.60 billion), contributing 40% to Australia's overall economic activity.[950] The estimated GDP per capita for Melbourne is A$72,664 (US$50,531) in the same year. Melbourne's economy was hit hard by the COVID-19 pandemic, primarily due to the prolonged lockdowns. Between March 2020 and October 2021, Melbourne experienced the longest lockdown in the world, lasting a total of 267 days.[951] This led to a recession in Melbourne during the 2020–2021 financial year, with a reduction of 1.6% in GDP.[952] In the 2021–2022 financial year, the city achieved a strong economic recovery, with a notable 5.4% GDP growth.[953] Key sectors driving this recovery included Construction, Health Care and Social Assistance, Retail Trade, Accommodation and Food Services, and Arts and Recreation Services.[954]

---

[947] State Government of Victoria. (n.d.). *Environment and Weather.* https://liveinmelbourne.vic.gov.au/live/environment-and-weather.

[948] Australian Bureau of Statistics. (n.d.a). *Greater Melbourne — 2021 Census All Persons QuickStats.* https://abs.gov.au/census/find-census-data/quickstats/2021/2GMEL.

[949] Australian Bureau of Statistics. (n.d.b). *Greater Sydney — 2021 Census All Persons QuickStats. https://abs.gov.au/census/find-census-data/quickstats/2021/1GSYD.*

[950] Wright, S. (2019). Melbourne Closes on Sydney as Nation's Economic Powerhouse. *The Sydney Morning Herald.* https://www.smh.com.au/politics/federal/melbourne-closes-on-sydney-as-nation-s-economic-powerhouse-20191215-p53k2h.html.

[951] Wahlquist, C. (2021). How Melbourne's 'Short, Sharp' Covid Lockdowns Became the Longest in the World. *The Guardian.* https://www.theguardian.com/australia-news/2021/oct/02/how-melbournes-short-sharp-covid-lockdowns-became-the-longest-in-the-world.

[952] SGS Economics & Planning. (2021). *National Economic Growth Hides the Differing Experiences Across Australia's Cities and Regions, New Report Reveals.* https://sgsep.com.au/publications/insights/australias-economic-wellbeing-2.

[953] SGS Economics & Planning. (2023). *SGS Cities and Regions Wellbeing Index.* https://sgsep.com.au/assets/main/SGS-Economics-and-Planning_CRWI.pdf.

[954] Ibid.

Regarding environment risks, while Melbourne's annual average temperature remained stable at 20.5°C across the past decades,[955] its annual average monthly rainfall varied significantly from 2002.[956] Melbourne suffered from the Millennium drought between 2001 and 2009, which led to the adoption of severe water restrictions.[957] In 2009, Melbourne had the lowest annual monthly rainfall (415 mm) and suffered a record-breaking prolonged heatwave over two distinct periods in January and February.[958] About 374 deaths were attributable to the 2009 extreme heatwave.[959] The Black Saturday bushfires, which occurred on and around 7 February 2009, resulted in the tragic loss of 173 lives in Victoria — the highest death toll from fire since colonisation in 1788. Additionally, 414 people were injured, and the fires scorched 450,000 hectares of land.[960] Prompted by these significant climate disasters, the City of Melbourne embarked on an ambitious pursuit of a holistic, cross-organisational agenda for environmental and climate resilience starting in 2010.[961]

The remainder of this chapter focuses on the City of Melbourne — a local government area located in the central city. It covers the central

---

[955] Australian Government Bureau of Meteorology. (2023a). *Mean Maximum Temperature*. http://www.bom.gov.au/jsp/ncc/cdio/weatherData/av?p_nccObsCode=36&p_display_type=dataFile&p_startYear=&p_c=&p_stn_num=086338.

[956] Australian Government Bureau of Meteorology. (2023b). *Monthly Rainfall*. http://www.bom.gov.au/jsp/ncc/cdio/weatherData/av?p_nccObsCode=139&p_display_type=dataFile&p_startYear=&p_c=&p_stn_num=086232.

[957] Moosavi, S., Browne, G. R., & Bush, J. (2021). Perceptions of nature-based solutions for Urban Water challenges: Insights from Australian researchers and practitioners. *Urban Forestry & Urban Greening*, *57*, 126937.; Bush, J., Oke, C., Dickey, A., Humphrey, J., Harrison, L., Amati, M., Fornari, G., Soanes, K., Callow, D., & Van der Ree, R. (2023). A decade of nature: Evolving approaches to Melbourne's 'nature in the city.' *Landscape and Urban Planning*, 235, 104754.

[958] *Ibid.*; McEvoy, D., Ahmed, I., & Mullett, J. (2012). The impact of the 2009 heat wave on Melbourne's critical infrastructure. *Local Environment*, *17*(8), 783–796.

[959] Cooper, M. (2009). Death Toll Soared During Victoria's Heatwave. *The Sydney Morning Herald*. https://www.smh.com.au/national/death-toll-soared-during-victorias-heatwave-20141112-9ubd.html.

[960] National Museum Australia. (2019). *Black Saturday Bushfires*. https://www.nma.gov.au/defining-moments/resources/black-saturday-bushfires.

[961] Bush *et al.* (2023). *Op. cit.*

business district (CBD) and 14 innermost suburbs, with a land area of around 37 km^2.[962] The City of Melbourne is governed by the Melbourne City Council, consisting of a lord mayor, a deputy lord mayor and nine councillors.[963] In 2022, the municipality had an estimated resident population of 159,813, with a population density of 4,281 persons per square kilometre.[964]

Over the past two decades, the City of Melbourne experienced a steady population growth until the onset of the COVID-19 pandemic in 2020.[965] Compared with the metropolitan area of Melbourne, the City of Melbourne has a younger population (median age 30 vs. 37) and a higher percentage of residents born overseas (55% vs. 36%).[966] This distinct population profile is largely due to the concentration of tertiary students, especially international students, who live on or around university campuses in the City of Melbourne, such as the University of Melbourne, RMIT, Monash University (Parkville Campus), and Victoria University. The population in the City of Melbourne is forecast to double and reach 308,000 by 2041, requiring over 42,000 new dwellings.[967]

Urban vibrancy in the City of Melbourne has seen a robust recovery from the pandemic. As of April 2023, morning commuting activity in the central city rebounded to 2019 levels on Tuesdays to Thursdays, driven by the ongoing popularity of flexible working arrangements. Consequently, the traditional Friday evenings of dining out and drinking with co-workers and friends shifted to Thursdays. Weekday pedestrian activity in the CBD returned to 72% of the 2019 level and 152% of the 2022 level. International students were back to 91% of the 2019 level, driving up more than 5% of spending in the central city.[968]

---

[962] City of Melbourne. (n.d.a). *About Melbourne*. https://www.melbourne.vic.gov.au/about-melbourne/Pages/about-melbourne.aspx.

[963] City of Melbourne. (n.d.b). *About Council*. https://www.melbourne.vic.gov.au/about-council/Pages/about-council.aspx.

[964] City of Melbourne. (n.d.c). *Community Profile*. https://profile.id.com.au/melbourne.

[965] *Ibid*.

[966] *Ibid*.

[967] City of Melbourne. (n.d.d). *City Forecasts*. https://www.melbourne.vic.gov.au/about-melbourne/research-and-statistics/Pages/city-forecasts.aspx?_ga=2.261457477.932182350.1656891285-1282394825.1622610528.

[968] City of Melbourne. (2023). *Melbourne Economy Snapshot: April 2023*. https://www.melbourne.vic.gov.au/SiteCollectionDocuments/melbourne-economy-snapshot-apr-2023.pdf.

11.2 Smart City Development: Smart Economy, Community, and Environment

The City of Melbourne's smart city vision is "to enhance the aspects of our city that make us uniquely Melbourne, and intelligently prepare for the changing needs of the community, the environment and the economy".[969] Smart city initiatives are coordinated by two teams within the Melbourne City Council, namely, the CityLab and the Emerging Technology Testbed. The CityLab is the innovation lab for the City of Melbourne, dedicated to prototyping and testing innovative ideas and council services in collaboration with the community. It takes a human-centric approach to recognise community needs and anticipate future challenges. The CityLab actively engages with the community to explore, build and test solutions to ensure that smart city initiatives are tailored to the residents.[970] The Emerging Technology Testbed serves as a collaborative space where the City of Melbourne partners with the community, industry, and businesses to explore the function of new and emerging technologies, such as 5G and IoT, and processes to support applications. It evaluates the impacts of new technologies at various scales, ranging from individuals, and communities, to the city.[971]

Smart city strategies have been developed in a series of strategic plans at the municipal level since the early 2000s, with the Council Plan 2021–2025 as the latest strategic policy guiding smart city development at the municipal level. More specifically, in the Council Plan 2004–2008, a key strategic goal for the City of Melbourne was to promote innovative urban development.[972] The concept of smart city development was first explicitly articulated in the Council Plan 2013–2017, seeking to harness technology

[969] City of Melbourne. (n.d.e) *Melbourne as a Smart City*. https://www.melbourne.vic.gov.au/about-melbourne/melbourne-profile/smart-city/Pages/smart-city.aspx?k=smart+city&type=web.

[970] City of Melbourne. (n.d.f) *CityLad*. https://www.melbourne.vic.gov.au/about-melbourne/melbourne-profile/smart-city/citylab/Pages/citylab.aspx.

[971] City of Melbourne. (n.d.g) *Emerging Technology Testbed*. https://www.melbourne.vic.gov.au/about-melbourne/melbourne-profile/smart-city/Pages/emerging-technology-testbeds.aspx.

[972] City of Melbourne. (2004). *Council Plan 2004–2008*. https://www.melbourne.vic.gov.au/SiteCollectionDocuments/council-plan-2004-08.pdf.

innovation to support strategic goals related to smart learning ('knowledge city' termed in the plan), smart mobility ('connected city'), and smart governance (open data, ICT infrastructure and applications in city service delivery).[973] More recently, the Council Plan 2021–2025 was launched in June 2021, one year after COVID-19, representing one of the major local policy responses for post-COVID recovery and long-term sustainable development. This plan aims to harness technology innovation in three strategic directions[974]:

- Smart economy.
- Smart community.
- Smart environment.

Adopting an integrated policy framework, smart city initiatives are further incorporated into a set of domain-specific strategies.

11.2.1 *Smart Economy*

The Council Plan 2021–2025 sets the priority to develop Melbourne as a digitally connected city that embraces new industries to create next-generation jobs and businesses. One major initiative is to collaborate with industry to support the development of globally competitive innovation ecosystems, including through international engagement, emerging technology trials and digital infrastructure delivery.[975] The plan also highlights the need for strengthening the creative sector, facilitating digital and technology innovation, and supporting the re-establishment of international education and efforts to unlock climate capital.[976] The council plan is further supported by the Economic Development Strategy 2031, which sets out a 10-year plan for economic, social, and cultural recovery for the

[973] City of Melbourne. (2013). *Council Plan 2013–2017*. https://www.melbourne.vic.gov.au/sitecollectiondocuments/council-plan-2013-17.pdf.
[974] City of Melbourne. (2021a). *Council Plan 2021–2025*. https://www.melbourne.vic.gov.au/SiteCollectionDocuments/council-plan-2021-25.pdf.
[975] *Ibid*.
[976] *Ibid*.

City of Melbourne as it emerges from the impacts of the COVID-19 pandemic.[977]

These smart economy strategies are implemented through a series of initiatives. The City of Melbourne supports start-ups and innovation businesses with co-working spaces, grants, and opportunities. The start-up community in Melbourne has access to an established, diverse, and educated workforce and world-leading universities.[978] Another example of the smart economy initiative is the pilot project of Micro-Labs: Reimagining Retail. This project develops a shared, co-working space that uses a variety of emerging technologies to explore ways to make retail spaces more desirable, feasible and viable for those returning to the city post-COVID. Technologies deployed include movement sensors, micro-climate sensors, air quality sensors, and interactive whiteboards and displays. Launched in June 2021, the pilot transitioned into a bookable space for events in March 2023 and concluded on 30 June 2023. Data collected at Micro-Labs are analysed to distil key insights on how people utilise the space and equipment and how to reshape traditional shopfronts to enhance support for the evolving needs of emerging occupations in the city's small businesses.[979]

11.2.2 *Smart Community*

The City of Melbourne has launched an open data platform, Open Data, to foster greater transparency and accountability.[980] This platform covers 224 datasets from 2019 to 2023 across eight themes: transportation, sensors, business, environment, community, property, city council, and events. Users can conduct analytics on the platform and visualise the

[977] City of Melbourne. (2021b). *Economic Development Strategy 2031*. https://www.melbourne.vic.gov.au/about-council/vision-goals/pages/economic-development-strategy-2031.aspx.

[978] City of Melbourne. (n.d.h). *Startups and innovation*. https://www.melbourne.vic.gov.au/business/start-a-business/startups/Pages/startups.aspx.

[979] City of Melbourne. (n.d.i). *Micro-Labs: Reimagining Retail*. https://participate.melbourne.vic.gov.au/emerging-tech-testbed/micro-labs?_ga=2.70035728.1224915442.1683512323-1646814621.1683097665.

[980] City of Melbourne. (2023a). *Open Data*. https://data.melbourne.vic.gov.au/pages/home/.

results.[981] The municipal government also provides free Wi-Fi across the city to allow residents and visitors to access free outdoor Wi-Fi services up to 1 GB per device, per day. Users do not need a personal login and there is no pop-up advertising.[982]

11.2.3 *Smart Environment*

The Council Plan 2021–2025 aims to enhance Melbourne's position as a global leader on climate action, showcasing world-leading sustainable design principles to enhance liveability and lead innovative responses to climate change. This includes transitioning towards zero net emissions by 2040, reducing waste through circular economies, mitigating public health impacts of climate change, and enhancing biodiversity, habitat, green spaces, water quality, and tree canopy cover in the city. Specific indicators are identified to evaluate the implementation progress towards greater environmental sustainability in the municipality.[983]

The environmental sustainability priorities outlined in the Council Plan will be delivered through a series of strategies, such as the Climate Change Mitigation Strategy to 2050 that aligns with the Paris Climate Agreement and the international effort to stay below a 1.5°C rise in global average temperatures.[984] One example of the smart environment initiative is the Cool Routes, a web-based mapping tool to help users find the coolest routes in central Melbourne. This tool integrates a variety of information, e.g., the day and time, shade, tree canopy and microclimate, to generate the coolest route possible between two points in the central area of the city. It may not provide the fastest walking/cycling route, but it identifies the coolest route possible to take advantage of

---

[981] City of Melbourne. (2023b). *Explore Our Data.* https://data.melbourne.vic.gov.au/explore/?sort=modified.

[982] City of Melbourne. (n.d.j) *Free wi-fi Across the City.* https://www.melbourne.vic.gov.au/about-melbourne/melbourne-profile/smart-city/Pages/free-wi-fi.aspx.

[983] City of Melbourne. (2021a). *Op. cit.*

[984] City of Melbourne. (2018). *Climate Change Mitigation Strategy to 2050.* https://www.melbourne.vic.gov.au/about-council/vision-goals/eco-city/pages/climate-change-mitigation-strategy.aspx.

existing shade and canopy cover to protect the user from the sun and the heat. This tool was identified as an action in the Climate Change Adaptation Strategy Refresh 2017, based on community concerns over safety and thermal comfort as Melbourne experiences more frequent, extended and intense heatwaves due to climate change. The mapping will be gradually extended to cover the whole of the City of Melbourne municipal area.[985]

Melbourne Smart City Timeline

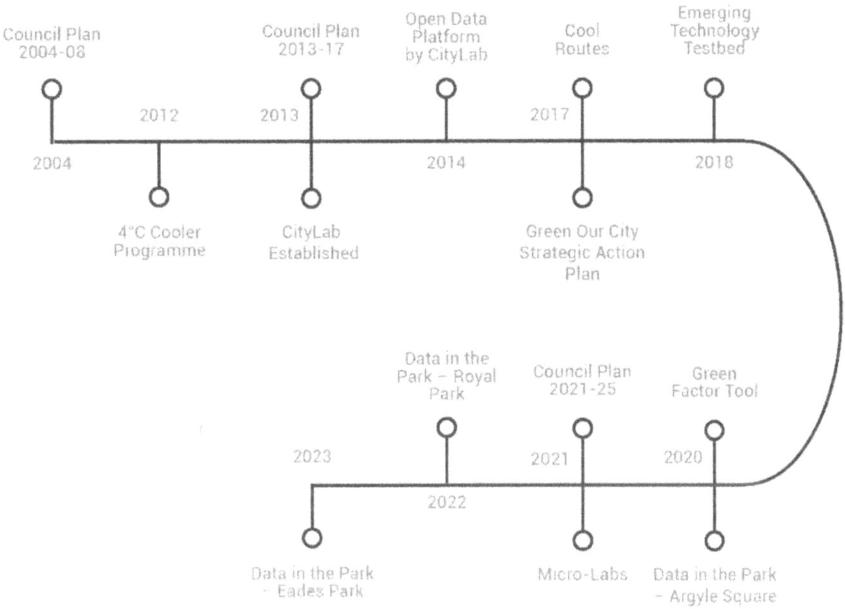

11.3 The 4°C Cooler Programme

The City of Melbourne launched the 4°C Cooler Programme in 2012 as an innovative approach to mitigating climate change impacts. Through strategically increasing urban forest and public open spaces across the municipality, the programme aims to increase the resilience of the urban forest to

[985] City of Melbourne. (n.d.k). *Cool Routes*. https://www.melbourne.vic.gov.au/parking-and-transport/streets-and-pedestrians/Pages/cool-routes.aspx.

cool Melbourne's summertime temperatures by 4°C.[986] The programme exemplifies the leadership role of the City of Melbourne in aligning with and supporting the broader environmental sustainability goals set for the Melbourne metropolitan area. These goals include creating "a greener city" (Strategic Direction 7), as established in the Melbourne 2030 Plan,[987] and the objective to "make Melbourne cooler and greener" (Strategic Direction 6.4), as outlined in the Plan Melbourne 2017–2050.[988]

The 4°C Cooler Programme was developed in response to the extreme heatwave and severe drought experienced between 1995 and 2009. Originally, the City of Melbourne aimed to reduce potable water usage by 90%, which involved a significant reduction in irrigation water supply to the urban forest and green spaces. Additionally, the city planned to remove 40% of trees throughout the municipality to lower water consumption. This policy response was soon recognised as problematic and short-sighted because it failed to recognise the crucial role of green infrastructure in mitigating the impacts of climate change. In 2010, the City of Melbourne appointed a new Urban Landscape Team to develop a long-term strategy to mitigate and adapt to climate change.[989]

After a two-year collaborative process of policy development with stakeholders, academics, and the community, the City of Melbourne endorsed the Urban Forest Strategy[990] and the Open Space Strategy[991]

---

[986] City of Melbourne. (n.d.l) *Cooling Our City*. https://www.melbourne.vic.gov.au/about-melbourne/sustainability/pages/cooling-our-city.aspx; UN Habitat. (2020). *4°C cooler — Using Green Infrastructure to Build a Climate Resilient and Prosperous Melbourne*. https://www.urbanagendaplatform.org/best-practice/4degc-cooler-using-green-infrastructure-build-climate-resilient-and-prosperous.

[987] City of Melbourne. (2002). *Melbourne 2030: Planning for Sustainable Growth*. https://www.planning.vic.gov.au/policy-and-strategy/planning-for-melbourne/melbournes-strategic-planning-history/melbourne-2030-planning-for-sustainable-growth.

[988] City of Melbourne. (2017). *Plan Melbourne 2017–2050*. https://www.planning.vic.gov.au/policy-and-strategy/planning-for-melbourne/plan-melbourne.

[989] UN Habitat. (2020). *Op. cit*.

[990] City of Melbourne. (2012a). *Urban Forest Strategy: Making a Greater City Greener 2012–2032*. https://www.melbourne.vic.gov.au/community/greening-the-city/urban-forest/pages/urban-forest-strategy.aspx.

[991] City of Melbourne. (2012b). *Open Space Strategy: Planning for Future Growth*. https://www.melbourne.vic.gov.au/community/parks-open-spaces/policies-plans/Pages/open-space-strategy.aspx.

in 2012. The overarching goal is to build "a city within a forest" to lower the city's temperatures. In the City of Melbourne, the urban forest consists of "all the trees and other vegetation — and the soil and water that supports it — within the municipality".[992] These guiding strategies seek to address three key challenges facing the City of Melbourne in the 2000s: (1) climate change impacts at the local scale (e.g., severe drought, water shortage, and more frequent extremes of heat and flooding mentioned previously), (2) rapid population growth from residents, workers, students, and visitors against the backdrop of urban intensification in the municipality, and (3) intensified urban heat island effect in the city.[993]

To guide the implementation of the 4°C Cooler Programme, the Urban Forest Strategy establishes a set of targets:

- To increase tree canopy cover on the public land across the municipality from 22% in 2012 to 40% by 2040.
- To increase urban forest diversity so that species composition is no more than 5% of any tree species, no more than 10% any genus and no more than 20% of any one family.
- To improve vegetation health to 90% by 2040.
- To improve soil moisture and water quality at levels to support the healthy growth of vegetation.
- To improve urban ecology and biodiversity to contribute to the delivery of healthy ecosystem services.
- To inform and consult with the community to ensure that the community will have a broader understanding of the importance of the urban forest, increase their connection to it and engage with its process of evolution.[994]

These targets are integrated into 10 urban forest precinct plans, which serve as guidelines for enhancing the urban forest in local neighbourhoods. Each precinct plan represents a 10-year planting strategy that assesses the existing conditions of the urban forest in the respective area (Figure 11.1). It identifies the necessary changes and outlines the steps to

[992] City of Melbourne. (2012a). *Op. cit.*
[993] *Ibid.*; City of Melbourne (2012b). *Op. cit.*
[994] City of Melbourne. (2012a). *Op. cit.*

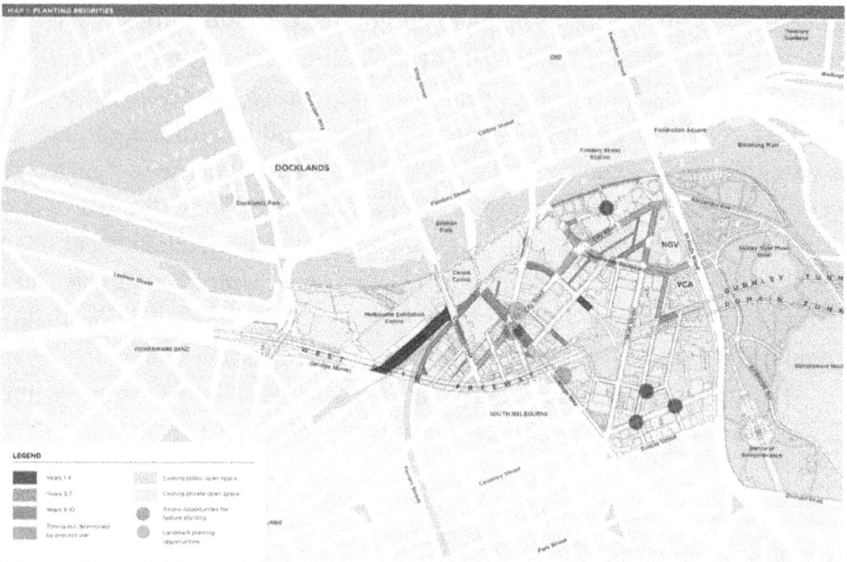

Figure 11.1. Southbank urban forest precinct plan 2015–2025.
Source: City of Melbourne.[995]

be taken to increase tree canopy area and create a healthier and more diverse urban forest.[996]

Furthermore, the Open Space Strategy seeks to provide a distributed open space network that is within easy walking distance for most of the community in the City of Melbourne. The municipality's public open space network comprises 148 sites, covering approximately 555 hectares, constituting nearly 15% of the total land area in the municipality.[997] The distributed open space network is expected to mitigate the urban heat island effect and provide environmental improvements in biodiversity, irrigation, and the treatment of urban runoff. This strategy also proposes additional and expanded open spaces, which are integrated into a series of open space precinct plans (Figure 11.2).[998]

[995] City of Melbourne. (2015). *Southbank Urban Forest Precinct Plan 2015–2015*. https://www.melbourne.vic.gov.au/SiteCollectionDocuments/ufpp-southbank-precinct.pdf.
[996] City of Melbourne (n.d.m) *Urban Forest Precinct Plans*. https://www.melbourne.vic.gov.au/community/greening-the-city/urban-forest/Pages/urban-forest-precinct-plans.aspx.
[997] City of Melbourne (2012b). *Op. cit.*
[998] *Ibid.*

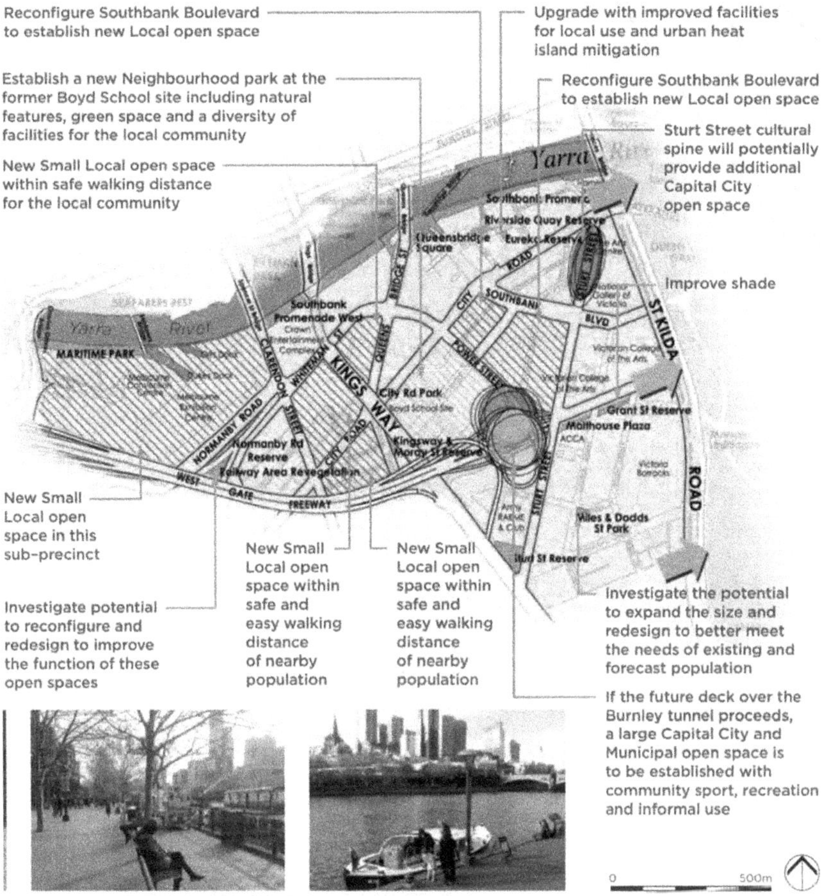

Figure 11.2. Southbank open space precinct plan.
Source: City of Melbourne.[999]

To facilitate effective implementation, the City of Melbourne highlights the importance of a long-term funding commitment from key stakeholders, including the public sector, developers, businesses, and the wider community.[1000] Within the public sector, the City of Melbourne secured investment of A$40 million from different levels of government as of

[999] *Ibid.*
[1000] City of Melbourne. (2012a). *Op. cit.*

2020, including the Council, the State, and the Federal government, to support the development of green infrastructure.[1001] The Open Space Strategy introduced a planning scheme amendment — 'Open Space Contribution Framework' — to require private sector developments to make financial contributions towards the provision of green infrastructure. These mandatory open space contributions are at rates of 8% and 5%, which can be fulfilled in the form of land or cash.[1002]

In addition, the City of Melbourne established the Urban Forest Fund in 2017 to accelerate greening across the city. This fund offers financial assistance for various new greening projects, encompassing the development of green spaces (Figure 11.3), tree planting initiatives, biodiversity projects, vertical greening, and the installation of green roofs. Individuals and organisations can apply for an Urban Forest Fund grant to create new green spaces on private property within the city. The fund will match each dollar contributed. Alternatively, individuals and organisations can

Figure 11.3. Urban green spaces in the City of Melbourne.

Source: Authors.

[1001] UN Habitat. (2020). *Op. cit.*

[1002] City of Melbourne. (2012b). *Op. cit.*

contribute to the fund to support greening efforts in the city.[1003] The Urban Forest Fund has provided four rounds of grants, totalling over A$1.7 million.[1004] The adoption of the Urban Forest Strategy, throughout the past decade, has served as a key funding facilitator for urban nature initiatives in the municipality, including cooling the city through the urban forest to other nature-based activities, e.g., orchard walks that demonstrate how the urban forest can contribute to food production.[1005]

Technology innovation underpins the 4°C Cooler Programme. Earlier efforts focused on measures to monitor and enhance the health of the urban forest. For instance, weather stations were installed throughout the CBD to monitor microclimate conditions at the streetscape level, considering different tree canopy compositions. Analytical tools like the 'i-Tree Eco tool' were utilised to calculate the monetary value of the environmental benefits provided by trees. Water-sensitive urban design features were implemented to support the development of a healthy urban forest, such as rain gardens, bioswales, underground storage reservoirs, and biofilters.[1006] In 2020, the City of Melbourne introduced the Urban Forest Visual platform, an open dataset and mapping tool as part of the Open Data platform, to enable urban forest data analytics and address some of the challenges facing the city's urban forest.[1007]

11.4 Implementation and Impact of the 4°C Cooler Programme

Over the past decade, the City of Melbourne has implemented the 4°C Cooler Programme through a range of innovative, nature-based solutions

[1003] City of Melbourne. (n.d.n). *Melbourne's Urban Forest Fund.* https://www.melbourne.vic.gov.au/community/greening-the-city/urban-forest-fund/Pages/urban-forest-fund.aspx.

[1004] City of Melbourne. (n.d.o). *Funded Projects.* https://www.melbourne.vic.gov.au/community/greening-the-city/urban-forest-fund/funded-projects/Pages/funded-projects.aspx.

[1005] Bush, J., Oke, C., Dickey, A., Humphrey, J., Harrison, L., Amati, M., Fornari, G., Soanes, K., Callow, D., & Van der Ree, R. (2023). A decade of nature: Evolving approaches to Melbourne's 'nature in the city.' *Landscape and Urban Planning*, 235, 104754.

[1006] City of Melbourne. (2012a). *Op. cit.*

[1007] City of Melbourne. (2020). *Urban Forest Visual.* http://melbourneurbanforestvisual.com.au/index.html.

for climate change mitigation and adaptation. This section elaborates on the implementation and impact of two of these innovations: Green Our City Strategic Action Plan and Argyle Square 'Smart Park' pilot project.

11.4.1 *Green Our City Strategic Action Plan*

Since its endorsement in 2012, the 4°C Cooler Programme has evolved from an initial focus on urban forestry to embracing a multi-faceted approach to city greening. This holistic approach is established in the Green Our City Strategic Action Plan, encompassing greening initiatives across various domains such as the implementation of green laneways, green roofs, and green walls and facades. This plan was launched in 2017 as the primary strategy for further greening the City of Melbourne, with a strong focus on delivering green infrastructure outcomes within the private realm.[1008]

More specifically, while the Urban Forest Strategy and Open Space Strategy provide a strong policy commitment to increase green infrastructure on public land, Melbourne City Council owns and controls less than one-third of the land area in the municipality. Given that 78% of the land in the municipality is privately owned or owned by other government entities, such as the Victoria State Government and the Federal Government, the City of Melbourne faces limited capacity to expand green infrastructure on private land. As of November 2017, there were only 38 green rooftops in the city, indicating a slow and minimal uptake of green infrastructure in the private sector. This constraint is further compounded by the absence of a mandatory requirement for green infrastructure in the local planning scheme.[1009]

To address these challenges, the Green Our City Strategic Action Plan was developed and implemented to guide greening initiatives across the following four focus areas[1010]:

[1008] City of Melbourne. (2017). *Green Our City Strategic Action Plan 2017–2021: Vertical and Rooftop Greening in Melbourne.* https://www.melbourne.vic.gov.au/community/greening-the-city/green-infrastructure/Pages/green-our-city-action-plan.aspx.
[1009] *Ibid.*
[1010] *Ibid.*

- **Leading by example**: This involves the implementation of research and demonstration projects to serve as models for further greening initiatives. One notable project is the Green Our Rooftop project,[1011] which is a collaborative effort between the City of Melbourne and the then Department of Environment, Land, Water, and Planning of the State Government of Victoria. The 1500-square-metre green roof at One Treasury Place (Figure 11.4) aims to showcase the successful retrofitting of buildings with high-quality green roofs through innovative research and deliver environmental and social benefits such as urban cooling, reducing energy consumption and reducing stormwater

Figure 11.4. Green our rooftop project at one treasury place.
Source: City of Melbourne.[1012]

[1011] Phillips, L. (28 June 2022). Interview with (Acting) Manager Open Space Planning and Green Infrastructure, the City of Melbourne [Personal communication].
[1012] City of Melbourne. (n.d.p). *Op. cit.*

runoff.[1013] Research grants and postgraduate scholarships totalling A$400,000 were collectively provided by the City of Melbourne, the Victorian Government, and Melbourne Water.[1014] These funds were allocated to support a diverse group of 13 researchers including PhD candidates and collaborative research projects. This initiative serves to foster innovation and develop co-funding and other funding arrangements.[1015]

- **Fostering partnerships**: The City of Melbourne collaborates with the Green Building Council of Australia to develop industry standards and rating tools for green buildings that not only address energy efficiency requirements but also incorporate green infrastructure considerations.[1016] In addition, the Urban Forest Fund mentioned earlier plays a key role in developing and maintaining partnerships with the private sector to co-fund green infrastructure projects within the private realm.[1017] Melbourne Skyfarm is a recent project that received an A$300,000 matched grant from the Urban Forest Fund.[1018] This project is a partnership between three Melbourne-based sustainability companies (Odonata, Biofilta, and the Sustainable Landscape Company) and the City of Melbourne to transform a 2000-square-metre rooftop of a multi-level car park into a productive and innovative urban farm on Siddeley Street, Docklands, in the central city.[1019] The urban farm features beautiful seating, a playground, vegetable and herb production, and fruit trees (Figure 11.5). It seeks to deliver multiple environmental benefits, such as microclimate cooling of the

[1013] City of Melbourne. (n.d.p). *Green Our Rooftop project.* https://www.melbourne.vic.gov.au/community/greening-the-city/green-infrastructure/pages/green-our-rooftop-project.aspx.
[1014] *Ibid.*
[1015] Phillips (28 June 2022). *Op. cit.*
[1016] City of Melbourne. (2017). *Op. cit.*
[1017] City of Melbourne. (n.d.p). *Op. cit.*
[1018] Phillips (28 June 2022). *Op. cit.*
[1019] City of Melbourne. (n.d.q). *Melbourne Skyfarm.* https://www.melbourne.vic.gov.au/community/greening-the-city/urban-forest-fund/funded-projects/pages/melbourne-skyfarm.aspx; Melbourne Skyfarm. (n.d.). *The Skyfarm Story.* https://www.melbourneskyfarm.com.au.

Figure 11.5. Melbourne Skyfarm.
Source: Melbourne Skyfarm.[1020]

rooftop, mitigating stormwater runoff, and reducing landfills by installing Biofilta food cubes made locally from 80% recycled food-grade plastic. Furthermore, the project seeks to provide community education programmes to enhance public awareness about the benefits of urban agriculture and sustainability. The construction of the Skyfarm was completed in December 2021 and is yet to open to the public.[1021]

- **Targeted advocacy**: Since 2008, the City of Melbourne has been delivering the Canopy Green Roof Forum on a quarterly basis to share industry best practices and to contribute to the capacity building of the development industry.[1022] This knowledge-sharing and networking

[1020] Melbourne Skyfarm. (n.d.). *Site Photos*. https://www.melbourneskyfarm.com.au/gallery.html.
[1021] City of Melbourne. (n.d.q). *Op. cit.*
[1022] City of Melbourne. (n.d.r). *Canopy Green Roof Forum*. https://www.melbourne.vic.gov.au/community/greening-the-city/green-infrastructure/Pages/canopy-green-roof-forum.aspx.

platform invites a variety of stakeholders from the industry, such as green infrastructure suppliers, designers, professional practice researchers, developers, and building owners, to share projects that have been delivered and to exchange learnings and approaches to overcome challenges in implementing green infrastructure projects.[1023]

- **Establishing effective regulations**: The Green Our City Strategic Action Plan recommends modifications to the local planning scheme to require all types of development within the municipality to actively contribute towards achieving environmentally sustainable design targets, including green roofs and vertical greening.[1024] In March 2023, the City of Melbourne proposed a Planning Scheme Amendment C376 Sustainable Building Design.[1025] The proposed amendment aims to establish a minimum mandatory requirement for all new developments to obtain a Green Factor score of 0.55. For alterations and additions exceeding 1,000 m^2 of gross floor area, a desired Green Factor score of 0.55 is targeted. To comply with this requirement, a Green Factor scorecard must be submitted as part of the landscape documentation for all relevant development applications. Public consultation regarding the amendment was conducted from 2 March to 17 April 2023 and the city council has been reviewing community feedback before a final decision is made.[1026]

The Green Factor score is generated by the Green Factor Tool, an award-winning online green infrastructure assessment tool, and is the first of its kind in Australia released in 2020.[1027] This tool was designed by the City of Melbourne in collaboration with the University of Melbourne, with the aim to support developers and designers to integrate more green infrastructure into their developments and measure the quality and quantity of those green infrastructure proposals. The tool calculates a Green

[1023] *Ibid.*; Phillips (28 June 2022). *Op. cit.*
[1024] *Ibid.*
[1025] City of Melbourne. (n.d.s). *Amendment C376: Sustainable Building Design.* https://participate.melbourne.vic.gov.au/amendment-c376.
[1026] *Ibid.*
[1027] City of Melbourne. (n.d.t). *Green Factor Tool.* https://www.melbourne.vic.gov.au/community/greening-the-city/green-infrastructure/Pages/green-factor-tool.aspx.

Factor Score to represent the proportional volume and effectiveness of green elements in relation to the overall site area. The tool can be used for evaluating and improving green infrastructure designs for new buildings and significant alterations and additions.[1028]

To supplement initiatives in these four focus areas, the City of Melbourne published a series of technical guides on the city council website to help the adoption of green infrastructure on private land. For instance, the Growing Green Guide provides homeowners, developers, and built environment professionals with comprehensive information about the design, planning, construction, installation, and maintenance of new green roofs, walls, and facades.[1029] The guide has gained recognition in the C40 Good Practice Guides, acknowledging its significant influence not only in promoting the adoption of green infrastructure within the City of Melbourne but also in inspiring other cities to find innovative solutions for addressing urban heat challenges. This technical guide is particularly valuable in situations where regulatory codes or public guidance and resources may be insufficient, but there is a clear public interest and willingness to explore alternative approaches like nature-based cool surfacing.[1030]

Within the context of Australia, while nature-based solutions are increasingly recognised for their potential to tackle climate change challenges, the adoption of urban policies and practices is still limited.[1031] Through increasing various forms of green infrastructure on both public and private land, the Green Our City Strategic Action Plan illustrates a multi-faceted approach to climate resilience that mitigates the urban heat challenge and promotes community health and well-being. This innovation serves as a valuable reference for other cities to develop smarter actions towards climate mitigation and adaptation. Globally, the impact of Melbourne's 4°C Cooler Programme has been recognised by several

[1028] *Ibid.*

[1029] City of Melbourne. (2014). *Growing Green Guide*. https://www.melbourne.vic.gov.au/community/greening-the-city/green-infrastructure/Pages/growing-green-guide.aspx.

[1030] C40 Cities. (2016). *C40 Good Practice Guides: Melbourne–Growing Green Guide*. https://www.c40.org/case-studies/c40-good-practice-guides-melbourne-growing-green-guide/.

[1031] Bush *et al.* (2023). *Op. cit.*

international organisations, such as being shortlisted for the 2nd Guangzhou International Award for Urban Innovation,[1032] part of the UN-Habitat best practice to build climate-resilient cities,[1033] and the C40 Good Practice Guides.[1034] As part of the Cool Cities Network within C40, the City of Melbourne demonstrates an early success story in tackling the urban heat challenge through green infrastructure.[1035]

Nevertheless, challenges may still arise during the ongoing implementation process including the following[1036]:

- **Cross-government collaboration in urban renewal areas**: In the City of Melbourne, the planning process for large-scale urban renewal precincts, such as Fishermans Bend, involves collaboration across various departments at the state government level, such as the Department of Planning, Department of Jobs, Precincts and Regions, and the Victorian Planning Authority. Additionally, the local government also works closely with agencies at the municipal level, such as Melbourne Water and Southeast Water, due to its involvement in water management and broader climate adaptation initiatives. This cross-government, multi-agency involvement presents both challenges and opportunities in the planning and implementation of green infrastructure projects, presenting a complex endeavour for pursuing greening initiatives in large-scale urban renewal areas.
- **Long-term and adaptive measures**: The City of Melbourne strives to ensure that the planning aligns with long-term goals for the metropolitan area of Melbourne, extending up to 2050, and effectively addresses climate risks. It is crucial to avoid maladaptive measures

---

1032 Guangzhou Awards. (2014). *Melbourne, Australia.* http://www.guangzhouaward.org/a/973.html?lang=en.

1033 UN Habitat. (2020). *Op. cit.*

1034 C40 Cities. (2016). *C40 Good Practice Guides: Melbourne–Growing Green Guide.* https://www.c40.org/case-studies/c40-good-practice-guides-melbourne-growing-green-guide/.

1035 C40 Cities. (n.d.). *Cool Cities Network.* https://www.c40.org/networks/cool-cities-network/.

1036 Phillips (28 June 2022). *Op. cit.*; City of Melbourne. (28 June 2022). Interview with the City of Melbourne [Personal communication].

and to design effective measures that mitigate the urban heat island effect and manage flooding risks.[1037]

Reflecting on the city's journey in urban greening and climate resilience, Melbourne City Council representatives identified several critical success factors[1038]:

- Establishing a clear set of strategic directions.
- Having a consistent framework to monitor and evaluate implementation progress.
- Maintaining funding stability from the public sector.
- Strengthening partnerships with the private sector for co-funding mechanisms.
- Pursuing changes to planning regulations.
- Fostering collaboration with local universities to strengthen the evidence base for green infrastructure innovations.
- Enhancing public awareness of the benefits of green infrastructure through education programmes and public engagement activities to ensure community buy-in and ownership.
- A dedicated team with the right expertise.

11.4.2 *Argyle Square 'Smart Park' Pilot*

In its pursuit of urban greening and cooling, the City of Melbourne also leverages smart technologies to enhance the monitoring, management, planning, and design of public open spaces in the municipality. Since late 2020, Argyle Square, near the University of Melbourne, has been serving as the living lab for a collaborative pilot project to use sensors and other smart devices to transform the public open space into a 'smart park'. A smart park is defined here as "a space that can express its environment and how it's used through data and technology."[1039] This project is a collaboration

---

[1037] *Ibid.*

[1038] Phillips (28 June 2022). *Op. cit.*; City of Melbourne (28 June 2022). *Op. cit.*

[1039] City of Melbourne. (n.d.u). *Data in the Park — Argyle Square.* https://participate.melbourne.vic.gov.au/emerging-tech-testbed/data-park.

Figure 11.6. Interactive map of sensors at Argyle Square.
Source: City of Melbourne.[1040]

between the City of Melbourne, the University of Melbourne and industry partners.[1041]

As the first open space sensing project in Melbourne, Data in the Park, Argyle Square pilot deploys seven types of sensors throughout the park to collect real-time data on microclimates, air quality, rubbish bin levels, pedestrian activity, and the usage of park stage and benches (Figure 11.6).[1042] Sensor data is integrated into the Open Data platform, mentioned earlier, to represent and analyse how humans and animals use park spaces. The evidence basis seeks to enable better management of parks in the future.[1043]

In relation to the 4°C Cooler Programme, the microclimate and tree canopy sensor data will help evaluate the effectiveness of the current approach to urban cooling through green infrastructure and increased tree canopy cover. These data strings will also help identify temperature differences between a healthy canopy location and a public location that had less or no canopy cover, enhancing the evidence basis for policymaking. Along with other sensor data on park usage patterns, the smart park pilot at Argyle Square seeks to understand what interventions help cool the city

[1040] *Ibid.*
[1041] *Ibid.*
[1042] *Ibid.*
[1043] City of Melbourne. (n.d.v). *Argyle Square Sensor Data Pilot.* https://melbournetestbed.opendatasoft.com/pages/overview-enhancement/.

Figure 11.7. Signages of public open space sensing at Argyle Square.
Source: Authors.

and simultaneously make public open spaces more comfortable for the community to use.[1044]

Another key feature of the Argyle Square smart park project is to empower the community to co-create future public spaces at Argyle Square. Over the past two years, the Emerging Technology Testbed team has conducted a series of community engagement activities to educate the residents about why and how sensor data will be collected and used and how the privacy of park users will be protected. "Privacy by design" is a key feature of the Argyle Square smart park project.[1045] Signages are provided to inform the public about what types of data are collected by sensors and that personal information is not collected (Figure 11.7).

Following the Argyle Square smart park pilot, public space sensing projects have been rolled out to additional parks in the City of Melbourne, including Royal Park,[1046] Eades Park,[1047] and Birrarung Marr.[1048] The goal

---

[1044] Phillips (28 June 2022). *Op. cit.*

[1045] Baxter, G. (11 August 2022). Interview with Pilot and Trial Lead (acting) City of Melbourne [Personal communication].

[1046] City of Melbourne. (n.d.w). *Data in the park — Royal Park.* https://participate.melbourne.vic.gov.au/emerging-tech-testbed/royal-park.

[1047] City of Melbourne. (n.d.x). *Data in the Park — Eades Park.* https://participate.melbourne.vic.gov.au/data-park-eades-park.

[1048] City of Melbourne. (n.d.y). *Data in the Park — Birrarung Marr.* https://participate.melbourne.vic.gov.au/emerging-tech-testbed/data-park-birrarung-marr.

is to further explore how sensors and other smart innovations can help better understand the usage patterns of public open spaces to inform park designs and management in the future.

Despite the scaling up of the smart park initiative, there are several ongoing challenges to overcome including the following[1049]:

- **Scale of efficiency**: Sensing open space will be substantially more useful if it is done at scale across a large number of open spaces. This will support the scale of efficiency, especially for operational tasks.
- **Communication with the public**: There is an ongoing effort to identify how best to inform the public about the value of public space sensing, what data will be collected by sensors and what signages/maps are needed. It is also important to clarify what data the sensors will not be collecting.
- **Data applications**: Open space sensing data needs to be communicated in a format that is meaningful for staff in other council departments so that the value of open space sensing could be optimised.

Critical success factors for the Argyle Square smart park project include the following[1050]:

- A problem-driven approach that goes by use case of sensors rather than by technology. This will ensure that sensor data is used to address real problems.
- User-centric and co-design from the start to bring along all stakeholders on board. This involves policymakers, planners, park operation teams, industry collaborators and the public.
- Sparking imagination with tangible examples of how sensor data can be used to address specific urban issues.
- Commitment of the project team that drives innovation towards tangible success.
- Technology and workforce capacity to ensure that the right sensors are installed and maintained and that sensor data can be integrated into a centralised, integrated data environment.

[1049] Baxter (11 August 2022). *Op. cit.*
[1050] *Ibid.*

11.5 Conclusion

Melbourne's smart climate action demonstrates an evolving, multi-faceted policy approach to innovation and transformation towards environmental sustainability. While the 4°C Cooler Programme was initially developed to mitigate the urban heat challenge, the guiding strategies have expanded from the initial focus on the urban forest and public open spaces to encompass various forms of green infrastructure in the private realm, such as green laneways, green roofs, green walls, and facades.

Periodic review and evaluation is essential. The launch of the Green Our City Strategy Action Plan in 2017 underscores the importance of periodic review and evaluation of urban greening policies to ensure that the long-term goal of cooling the city through green infrastructure can be effectively implemented. Demonstration projects, such as the Green Our Rooftop project and the Melbourne Skyfarm project, provide tangible examples of co-funding mechanisms and leveraging technology innovations to further green the city:

Leveraging smart technologies: Technology innovation plays a critical role in the implementation of the 4°C Cooler Programme. Collaborative research projects have been carried out by the City of Melbourne and local universities to strengthen the knowledge basis of green infrastructure interventions for cooling the city. Digital tools and platforms, such as the Open Data platform and the Green Factor Tool, enable multiple stakeholders, such as developers, owners, designers, and other built environment industry practitioners, to make joint efforts in increasing green roofs and vertical greening in the city, particularly within the private realm.

Advocating for regulatory changes: The Open Space Contribution Framework and the recently proposed Planning Scheme Amendment C376 Sustainable Building Design provide examples of how the local government embed green infrastructure requirements in the local planning regulations. These regulatory changes seek to support a more effective implementation of urban greening across the municipality.

Engaging and co-creating with the community: Community engagement, especially in the Urban Forest Strategy, fostered community interest in urban nature-based solutions for addressing climate change impacts.[1051] In the more recent initiative of Smart Park pilots, community engagement activities were organised to inform the public about the benefits of public space sensing and to empower the community to participate in the future designs of parks. It also reassures the public that personal information is not collected by sensors in smart parks to ensure data privacy.

[1051] Bush *et al.* (2023). *Op. cit.*

CHAPTER 12

MILAN: SUSTAINABLE FOOD AVAILABILITY

12.1 City Overview

Location	Milan — Capital City of Lombardy, Italy.
Land Size[1052]	182.1 km^2
Climate[1053]	Continental
Geography (topology/ landscape)[1054]	Arid plains and marshland
Environmental Risks[1055]	Air pollution
Population[1056]	1,371,498 (2022)
GDP[1057]	US$87.7 billion (2019)
GDP Per Capita[1058]	US$63,944 (2019)
Key Urban Issues of Smart City	Food inequity, siloed government departments, unsustainable waste management, obesity, and malnutrition

[1052] Foot & Lecco (2023). *Op. cit.*
[1053] *Ibid.*
[1054] *Ibid.*
[1055] Legambiente (22 January 2019). *Op. cit.*
[1056] citypopulation.de (n.d.). *Op. cit.*
[1057] C40 Cities. (2019). *Milan.* https://www.c40.org/cities/milan/.
[1058] *Ibid.*

Milan is the capital of the region of Lombardy in northern Italy and the leading financial, manufacturing, and commercial city in Italy, spanning 182.1 km^2.[1059] The region is bordered on the north by Switzerland and the Italian regions of Emilia-Romagna (south), Trentino-Alto Adige and Veneto (east), and Piedmont (west). Milan lies within the Po River basin and has a continental climate with damp, cold winters and hot, humid summers. There is usually snowfall between December and February with temperatures ranging from −1°C to 10°C in winter and 20–30°C in summer. Fog often shrouds the city in winter, and the city faces environmental hazards of air pollution from urban pollutants and poor air circulation due to its geographical location and urban density.[1060]

In 2022, Milan had a population of 1,371,498, and a population density of 7,677 residents per square kilometre. 81.8% (1,149,794) of the residents in Milan are Italian citizens, 104,245 residents have Asian citizenship, 57,713 African citizenship, and 29,921 European Union citizenship, followed by 20,854 from the rest of Europe. The remaining 43,715 are from the rest of the world.[1061] 22.6% of the population residing in Milan are aged 65 and above with the Gallaratese neighbourhoods being identified as an area with the highest concentration of senior residents.[1062] Milan also faces the issue of low birth rates, with the number of births reaching a record low at 9,700 in 2019 — the lowest in a century. However, Milan sees its population growing due to high levels of local and international migration due to its vibrant economy.[1063]

[1059] citypopulation.de. (n.d.). Milano (Metropolitan City, Lombardia, Italy) — Population Statistics, Charts, Map and Location. https://www.citypopulation.de/en/italy/lombardia/015__milano/; Foot, J. & Lecco, A. (2023). Milan — Landscape | Britannica. https://www.britannica.com/place/Milan-Italy.

[1060] *Ibid.*; Legambiente. (22 January 2019). Mal'aria 2019, il rapporto annuale sull'inquinamento atmosferico nelle città italiane • Legambiente. Legambiente. https://www.legambiente.it/comunicati-stampa/malaria-2019-il-rapporto-annuale-annuale-sullinquinamento-atmosferico-nelle-citta-italiane/.

[1061] citypopulation.de (n.d.). *Op. cit.*

[1062] ARUP. (2018). *Shaping Ageing Neighbourhoods.* https://www.arup.com/en/perspectives/publications/promotional-materials/section/shaping-ageing-neighbourhoods-milan-gallaratese.

[1063] The Local Italy. (13 January 2020). *Population of Milan Hits 30-year High After Influx of New Residents.* https://www.thelocal.it/20200113/population-of-milan-hits-30-year-high-after-influx-of-new-residents/.

Milan resides within the region of Lombardy, which has an elected regional government headquartered in Milan. They oversee the management of regional administration and public health. The City Council of Milan (Italian: Consiglio Comunale di Milano) is the top-tier administrative body which consists of the elected Mayor of Milan and the council. This body governs the city's large-scale urban activities, such as public education, traffic management, public infrastructure, and public transportation. The province of Milan (Italian: Provincia di Milano) — succeeded by the metropolitan city of Milan (Italian: Citta Metropolitana di Milano), governs the area around the city, leading infrastructural development and cultural policies.[1064]

Milan is Italy's economic and financial centre, generating the highest GDP per capita in Italy.[1065] Milan's economy is anchored in fashion, manufacturing, banking, tourism, and the medical industry[1066]; 5% of Italian and 35% of Lombardy companies are in Milan. These companies produce approximately 9% of the added value of the Italian industry and export 10% of national exports (about €33,000 million). Milan hosts over 40 foreign banks, major fashion companies (e.g., Armani, Versace, Prada, Gucci, and Valentino headquarters), and major manufacturers in the automobile industry (Alfa Romeo and Pirelli). Other Milanese products include chemicals, machinery, pharmaceuticals, and plastics. Milan generates 9% of the national GDP while the Lombardy region generates 20% of Italy's GDP.[1067]

12.2 Smart City Development: Improving New Urban Economies and Social Inclusion

Since 2011, the municipality of Milan has been working on a series of innovations to improve new urban economies and social inclusion

[1064] Foot & Lecco (2023). *Op. cit.*

[1065] OECD. (2022). OECD Regions and Cities at a Glance 2022. OECD. https://doi.org/10.1787/14108660-en.

[1066] International Society for Magnetic Resonance in Medicine, & The European Society for Magnetic Resonance in Medicine and Biology. (2014). Milan: A City on the Move. International Society for Magnetic Resonance in Medicine, The European Society for Magnetic Resonance in Medicine and Biology. https://www.ismrm.org/14/tech/milan.pdf.

[1067] International Society for Magnetic Resonance in Medicine & The European Society for Magnetic Resonance in Medicine and Biology (2014). *Op. cit.*

Figure 12.1. Five lines of action in Milan for innovation.
Source: Municipality of Milan — Urban economy and employment.[1068]

Figure 12.2. Seven smart city thematic working groups.
Source: Municipality of Milan — Urban economy and employment.[1069]

in the most fragile areas of the city. This follows five lines of action (Figure 12.1), spanning three sectors: participatory planning, impact measurement, and impact financing. Milano Smart City project and public consultations began in December 2012, kickstarting the city's adoption of policy plans through urban development, sustainable mobility, energy efficiency and sharing economies.[1070] April 2013 marked the first public hearing of this proposal by Milano Smart City in collaboration with the Chamber of Commerce, and the city had public consultations with different stakeholders within the municipality and the territory up till May 2014 to create seven smart thematic working groups (Figure 12.2).[1071]

In 2016, Milan participated in the Sharing Cities Lighthouse Programme, activating international collaboration between industries and cities to develop integrated smart city solutions. This showcased the effectiveness of new technologies implemented to improve urban mobility,

[1068] *Ibid.*
[1069] *Ibid.*
[1070] European Commission (n.d.). *Milan | Intelligent Cities Challenge.* https://www.intelligentcitieschallenge.eu/cities/milan.
[1071] Municipality of Milan — Urban Economy and Employment. (2019). *Milan Strategy for Smart City: Technologies Enabling Economic Boost and Social Inclusion.*

energy efficiency, and reduction of carbon emissions in the Porta Romana/Vettaabbia district. This project connected the city's historic centre to its agricultural belt by introducing a functional mix of private and social housing units, new sustainable transport systems (e-cars, e-bike charging points, and smart parking), a new public transport station, and a large park. This project was funded by the European Union Horizon 2020 research and innovation programme.[1072] Since then, the municipality has established key milestones in supporting and financing the following new initiatives:

- **Hybrid enterprises and urban regeneration**[1073]: This supports bottom-up hybrid enterprises with social impact that blends profit and non-profit forms by combining trade, agriculture, food, volunteering, art, and cultural fields. 2013–2018 saw the establishment of new policies for hybrid economies through microcredit, reusing public buildings and funding for social enterprises. In 2016, Base Milano was established through the regeneration of old industrial spaces to become a cultural hub open to events, co-working spaces, and creative initiatives. The first civic crowdfunding began in 2016–2017 to fund social impact civic projects; 50% of the budget would be through crowdfunding, while the other 50% was met by public administration. From 2017 to 2019, the Open Agri Project was established. This was an experimental urban agriculture district in the south of Milan. Most recently, from 2019 to 2022, the Scuola dei Quartieri (School of the Suburbs) was established as a municipal project to increase the planning capacity of its citizens. Its goal is to generate inclusive citizen feedback for urgent improvements in its neighbourhoods.[1074] These initiatives use technology to increase efficiency and collaboration, driving the urban regeneration processes and creating social cohesion.

[1072] European Commission (n.d.). *Op. cit.*

[1073] Municipality of Milan — Urban Economy and Employment (2019). *Op. cit.*

[1074] The School of the Suburbs — Citizens Make the City. (n.d.). *The Innovation in Politics Institute.* https://innovationinpolitics.eu/showroom/project/the-school-of-the-suburbs-citizens-make-the-city/.

- **Sharing and collaborative economy**[1075]: The Sharing City ecosystem involves different actors participating in a process of co-design, co-development, and co-management of practices, spaces, goods, and services. There are various operators and experts guiding the Milan sharing and collaborative economy, ranging from co-working spaces that have been funded and active since 2012 to different vehicle-sharing platforms and initiatives (e.g., bicycles to e-scooters, cars, and electric vehicle charging points). To successfully establish this, mapping, enhancing, connecting, and regulating territorial initiatives linked to sharing and collaboration is necessary.
- **Start-up and knowledge-intensive economy**[1076]: This generates work, especially for Milan's younger population. This is achieved through investment in new strategic clusters to create innovative entrepreneurship in synergy with universities, research centres, and the third sector. These initiatives are supported by the Chamber of Commerce and experiments with innovative services to increase social impact in the peripheries. Since 2013, there have been ongoing policies to support start-ups in partnership with entrepreneurs' organisations and the Chamber of Commerce. Fabriq (established in 2014) was the first incubator for social start-ups and grants. There were subsequent international exchanges with agreements in 2016 with Amsterdam and New York City for study visits. In 2017, the European scaling-up network was started to support the growth of start-ups. Additional support was given to social impact start-ups in 2019, which supported experimentation of innovative services.
- **New craft and urban manufacturing**[1077]: Milan plans to create an enabling ecosystem for creating, establishing, and growing digital manufacturing and new crafts companies to generate new jobs, regenerate the suburbs, and promote social cohesion. Policies for fab labs and maker spaces began in 2013 in the form of investment support, collaboration for educational projects, and the creation of an official register. From 2016 to 2017, the Action Plan Codesign was a

---

[1075] Municipality of Milan — Urban Economy and Employment (2019). *Op. cit.*

[1076] *Ibid.*

[1077] *Ibid.*

participatory process of co-design sessions with more than 50 stakeholders to develop this ecosystem, the Manifattura Milano Plan provided directions for developing new craftsmanship and digital manufacturing from 2016 to 2020, and the Milano Luiss Hub was opened in 2017 as a new space for private-public partnerships for makers and students. This was followed by the Manifattura CAMP in 2018, which is a platform that matches players in manufacturing 4.0 with makers, universities, and research centres.

- **Smart city and smart citizens**[1078]: Milan's smart city plan innovates for tangible and intangible infrastructures through the regeneration of urban spaces, rethinking economies, and creating dialogues to manage complexity. Intangible processes such as maintaining social cohesion, changing behavioural habits, training, research, and participation

Milan Smart City Timeline

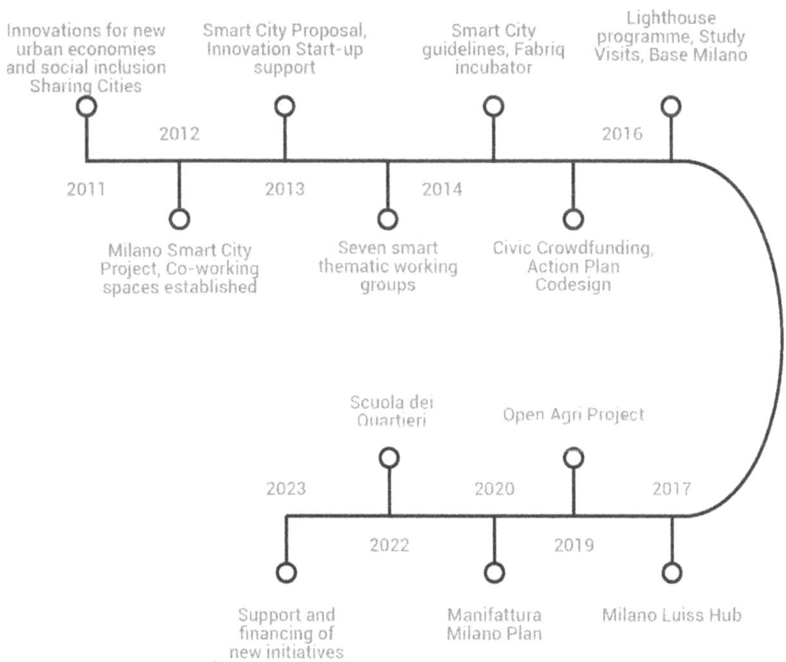

[1078] *Ibid.*

are equally important. The city provides free Wi-Fi, open datasets for citizens and businesses, and sensors to collect information on tourism and mobility. Sharing Cities Project Horizon 2020 is a major smart city project with 25,000 m^2 of private and public buildings undergoing energy redevelopments. New mobility provisions include bike stations, electric charging, shared electric vehicles, smart parking stalls, and electric vehicles for goods logistics.

12.3 Urban Food Policy

One of Milan's key smart city missions is to tackle the significant social and environmental issues caused by urban food waste. International food loss and waste emit around 4.4 billion tonnes of CO_2 every year. Milan plans to address this issue through greater redistribution of food losses across private and public sectors, business waste tax reduction to incentivise surplus food donations, greater local and international outreach on food waste, and transition towards a circular economy.[1079]

This initiative began in 2015 during Expo Milano, "Feeding the Planet, Energy for Life"[1080] which consolidated international policies, such as the United Nations Sustainable Design Goals (SDGs), especially SDG 2 (Zero Hunger), 11 (Sustainable Cities and Communities), 12 (Responsible Consumption and Production), 13 (Climate Action), and 16 (Peace, Justice, and Strong Institutions), and the Zero Hunger Challenge launched in 2012. The Milan City Council voted to adopt the "Milan Food Policy Guidelines", which established five priorities of the municipal food strategy[1081]:

- Provision of healthy food and water for all citizens
    - Ensure access to healthy drinking water and sufficient food as primary nourishment to protect human dignity and improve the quality of life.

[1079] Gordon, M. & Bidault, M. (2019). Cities 100 Report. C40 Cities, Realdania, Nordic Sustainability.

[1080] Gavazzeni, F. (7 June 2022). Interview on the Milan Urban Food Policy Pact [Personal communication].

[1081] Food Policy di Milano. (2015). *Objectives and Priorities — Food Policy di Milano.* https://foodpolicymilano.org/en/objectives-and-priorities/.

- Promote the sustainability of the food system
  - o Facilitate the consolidation of all the components and activities necessary for managing a sustainable food system and promote local production and consumption of fresh and seasonal quality food.
- Promote food education
  - o Promote a culture oriented to consumer awareness of healthy, safe, culturally appropriate, and sustainable food, produced and distributed with respect for human rights and the environment.
- Fight against food waste
  - o Reduce surpluses and food waste during the different stages of the food chain as a tool for limiting environmental impact and to contrast social and economic inequalities.
- Support scientific research in the agri-food sector
  - o Fostering the development of Milanese agri-food scientific research with connotations or implications related to the urban system, aiming at improving processes, and developing cutting-edge technologies.

Resultingly, The City of Milan developed the Milan Food Policy in partnership with the Cariplo Foundation after signing a Memorandum of Understanding in 2014 to implement and promote the initiative. The former mayor of Milan had the vision to create momentum through a platform for a city-to-city exchange on sustainable development with the entry point of food. This initiative is being managed by two departments within the Comune di Milano — the Milan Food Policy (MFP), which focuses on local food sustainability measures, research and local sustainability outreach, and the Milan Urban Food Policy Pact (MUFPP), which works on international outreach and collaboration on improving international urban food waste sustainability. Food policies were established by MFP to make Milan more inclusive and resilient, while MUFPP facilitated international exchange to accelerate innovation and development.[1082]

Milan aims to reach the goal of reducing food waste by 50% by 2030 with the help of local actors (stakeholders and shareholders). Through the MFP implementation, the city will try to address all these actors to achieve this goal. The policies for the reduction of food losses and waste

[1082] *Ibid.*

are combined with those related to waste cycle management to increase the overall sustainability of the system with an approach to a circular economy. These policies include the following[1083]:

- To inform and educate citizens and local stakeholders on reducing food losses and waste.
- To recover and redistribute food waste.
- To create local partnerships such as among charities food banks, supermarkets and municipal agencies.
- To improve and reduce food packaging.
- To strive for a circular economy in food system management.

12.4 Implementation and Impact of Milan Food Policy

This section elaborates on the implementation and impact of two of these innovations: the Milan Food Policy (MFP) and the Milan Urban Food Policy Pact (MUFPP).

12.4.1 *Milan Food Policy*

The Milan Food Policy Office was established in 2017 and has since engaged Milan's many assets in the city and rural agricultural regions. The MFP Office now runs services to promote sustainable food production and collaborates with local, regional, and country-level governments to establish food sustainability-related innovations. The MFP acts on the urban food system around five key priorities: food access, education, food production, food waste management, and scientific research.[1084] MFP also engages several local actors, such as research centres, private sectors, non-profit organisations, and foundations.[1085] This follows the UN SDGs,

[1083] C40 Cities (2019) *Op. cit.*

[1084] Porreca, E. (9 June 2022). Interview with Milan Food Policy [Personal communication].

[1085] C40 Cities. (2019). *C40 Cities Annual Report 2019.* C40 Cities Climate Leadership Group, Inc. https://www.c40.org/wp-content/uploads/2021/11/C40-2019-Annual-Report.pdf.

a main political commitment of Mayor Scala, and is supported by Vice Mayor Anna Scavuzzo. Milan also has inter-sectoral collaborations with Cariplo Foundation to transition to a more resilient food system.[1086]

12.4.2 *The Food Waste Hub*

In 2018, the municipality commissioned a study by Politecnico di Milano and Assolombarda, an association bringing together supermarkets, businesses, and company canteens, to design a pilot to collect and redistribute food waste on a community level. Banco Alimentare, the charity which presently runs the hub, started in 2008 and collected food from supermarkets for their food bank. The food hub (Figures 12.3 and 12.4) began development in 2015 when the charity met with the Assolombarda union and Politecnico di Milano during the Milan expo. They collaborated to establish a dedicated neighbourhood hub, where Banco Alimentare translated their already established Food Bank systems and network into the operations of the food hub.[1087]

Figure 12.3. Opening of a neighbourhood food hub.
Source: Politecnico Milano 1863.[1088]

[1086] C40 Cities Climate Leadership Group, Nordic Sustainability. (October 2019). *Op. cit.*
[1087] Cosentino (6 June 2022). *Op. cit.*
[1088] Politecnico Milano 1863. (5 February 2021). Fighting food waste: The success of the neighbourhood Hubs for recovery and support for the most vulnerable. *Politecnico Di Milano School of Management.* https://www.som.polimi.it/en/fighting-food-waste-the-success-of-the-neighbourhood-hubs-for-recovery-and-support-for-the-most-vulnerable/.

Figure 12.4. Entrance of food waste hub, Isola District.
Source: Authors.

The first pilot hub was set up in the Isola neighbourhood in January 2019 and was successful, leading to the opening of two new hubs in other parts of the city. The food waste hubs, located in municipality-owned warehouses, are run by community volunteers as part of the charity. They target to recover around 70 tonnes of food waste per year from supermarkets and company canteens and provide it to charities that organise distribution to beneficiaries.

To overcome food poverty and improve overall sustainability, the hub coordinates the logistics of food donations between different network actors, connecting 300 charities and hundreds of thousands of people to food sources. It aims to address the needs of people who are unable to rely directly on the municipality to support their food needs. The hub operates five days a week with collection from six different retailers in twelve locations in the morning. The food is then sorted for quality and collected by charity organisations in the afternoon to serve at soup kitchens (Figure 12.5).

Banco Alimentare manages the data of charities involved, food supplied, and the amount collected and distributed and handles schedules and

Figure 12.5. Food to be distributed to charities.
Source: Authors.

problems which may arise from the food collection process. Organisation and training of volunteers are provided by Banco Alimentare to ensure the safety of food transportation, preservation, and product handling. Food is also screened for health and safety, along with their perishability. Success is measured through KPIs established based on food collection and distribution quantities, and their growth through the years. Social factors like the satisfaction of charities and the ability to feed people in need are also equally important. Key support and resources are funded by private and public donations. Operations require €60,000 a year for facility costs, utility, and pay. This covers operating costs and infrastructural overheads such as transportation vehicles (Figure 12.6), gas, hired drivers, and storage equipment that keeps the collected food fresh. Each hub needs about 10–15 volunteers to ensure smooth operation. Volunteers play a large role in assisting with the food collection, sorting, data management, and logistics tracking process.

This initiative has been particularly impactful and it was measured in 2021 that the food hub collected 100 tonnes of food, corresponding to an

Figure 12.6. Surplus food collection van.
Source: Authors.

economic value of around €400,000 a year. The hub has large-scale economic negation based on the cost–benefit analysis (social economic and environmental value) and reducing overall CO_2 wastage if the food were thrown away.[1089] Banco Alimentare also works on other food waste collection systems, transporting surplus food from markets and canteens to their 3,000 m^2 logistics centre to the North of Milan, distributing over 20,000 tonnes of food in 2021. Food collected are raw bread and fruit from school canteens, already cooked but unserved meals, excess warehouse goods (cold storage), and fresh food (vegetables, cheese, eggs, and dairy) from local supermarket surplus (Figure 12.7).[1090]

The Food Waste Hub provides a transferrable model to other cities as its transparent practices and simple framework are easily reproduced. In Milan, more hubs have been planned for opening with the goal of having one hub in every district, providing comprehensive coverage of neighbourhood charity organisations and soup kitchens. To do this, a formal monitoring system needs to be set up to expand their logistics operations and outsource them to manage other charities and regions. KPIs and data must also be collected and processed to prove the success of the model for wider-scale adoption.[1091] To successfully scale this model, the following is needed:

[1089] Cosentino (6 June 2022). *Op. cit.*
[1090] *Ibid.*
[1091] Cosentino (6 June 2022). *Op. cit.*

Figure 12.7. Banco Alimentare surplus food collection from foody wholesale market.
Source: Authors.

- Trust and social capital in operating neighbourhoods and collaborators.[1092]
- Manpower with a core team allocated to organise and run operations full time,[1093] and a competent team that is passionate about the cause.[1094]
- Food management and quality control to ensure health and safety.[1095]
- Well-planned locations near neighbourhoods in need.[1096]
- Easily accessible with sufficient resources (quality supermarkets, charity resources, and volunteers).[1097]

[1092] *Ibid.*
[1093] *Ibid.*
[1094] *Ibid.*
[1095] *Ibid.*
[1096] *Ibid.*
[1097] *Ibid.*

12.4.3 *Milano Ristorazione*

Milano Ristorazione began in 2000 and is a Municipal Agency for School Canteens, 99% owned by the City of Milan. They have 24 outlets in the city which serve schools in different districts (Figure 12.8). It offers healthy, good, educational, and fair meals to children from kindergartens to middle schools in municipality schools, private schools, and summer camps. At present, this programme is partially covered by the municipality, and citizens pay a nominal fee for the food based on household income. The food production centres need to be certified for food quality, food safety management, agri-food tracing system, and environmental management.[1098]

Milano Ristorazione continues a legacy of social welfare that began 122 years ago in 1900. The municipality established this food service to cater to an impoverished population at the time. Post-war conditions made mandatory elementary schooling challenging as families struggled with affording food. Therefore, free meals were provided in school as a social welfare programme, simultaneously encouraging elementary education.

Figure 12.8. Milano Ristorazione preparation facility.
Source: Milano Ristorazione.[1099]

[1098] Milano Ristorazione. (9 June 2022). Tour of Milano Ristorazione Kitchen and Logistics Hub.

[1099] Milano Ristorazione (Director). (2 March 2022). *Centro Cucina di via Quaranta: La struttura e la tecnologia*. https://www.youtube.com/watch?v=ZXYI9vwW6LI.

Figure 12.9. Food preparation and packing area.
Source: Milano Ristorazione.[1100]

At present, food is provided for low-income families at highly subsidised rates or for free. Another socio-cultural factor for the continued legacy is the importance of food in Italian culture as a means of social integration, especially for the immigrant community.[1101]

At the food production factory, there are rooms dedicated to preparing various types of diets, which are individually packed (Figure 12.9). Meals are labelled with information on ingredients, types of dishes, and dietary requirements which accounts for health restrictions and religious diets. Special meals are prepared separately to prevent cross-contamination (Figure 12.10). Food is prepared on the day of serving, with machines to keep the cooked food at a regulated temperature — subjected to the types of dishes that day. Pasta, for example, is sent out immediately to maintain the quality and texture of the pasta. Styrofoam boxes maintain the temperature of meals and are then brought out in vans to individual schools. These are reusable and brought back from the schools to the centre every day.[1102]

[1100] Milano Ristorazione (Director) (9 June 2022). *Op. cit.*
[1101] Notarangelo, B. (9 June 2022). Interview with Milano Ristorazione [Personal communication].
[1102] Milano Ristorazione (9 June 2022). *Op. cit.*

Figure 12.10. Cooking and warming stations.
Source: Milano Ristorazione.[1103]

MFP works with Milano Ristoazione to incorporate food sustainability into its operations. This includes sustainable menus, food education, nudging behaviour to change unhealthy dietary habits, and other sustainable practices.[1104] EU also funds certain educational initiatives. This has led to the ability for them to practice more sustainable forms of production — such as using self-generated energy and having in house water treatment to remove grease before releasing it back into the municipal sewage system.

Such initiatives have been impactful in the goal to reduce urban food waste. As part of their nudging practices, Milano Ristoazione has cut down on red meat and removed deep-fried food while introducing more plant-based diets. They have reduced CO_2 by 20% in 5 years by introducing more vegetables, preferring steam cooking methods, transitioning to reusable cutlery, and using municipal tap water in refillable bottles. There are also programmes and brochures with different activities for age groups to educate them on the reduction of food waste. One pilot programme introduced doggy bags for students to put their leftover fruit and bread to

[1103] *Ibid.*
[1104] *Ibid.*

take home.[1105] Overall, the campaign resulted in a 17% decrease in food waste, with the participation of 57 schools covering 19,000 children (28% of the total). Together with Banco Alimentare, Milano Ristorazione also connected 106 schools to donate surplus food to charities.[1106]

Milano Ristorazione serves as a good example of positive influence which can be made by large-scale organisations in improving the sustainability of food production and procurement practices through the supply chain. By encouraging local procurement of organic and sustainably produced food, they improve public health and local farming practices which both stimulate the local agricultural economy while encouraging suppliers to produce food in healthy and sustainable ways.[1107] In scaling up sustainable practices, Milano Ristorazione also highlights the importance of maintaining the quality of services provided through external reviews of their industrial practices and opening up dialogues for community involvement in their operations (such as reviewing school menus and dietary provisions).[1108] This ensures that despite large-scale implementations, they remain grounded and responsive to community needs while ensuring local knowledge and support for their initiatives. However, due to the scale of their operations, this also means that sustainable changes take more time and effort to implement. One such example would be additional costs borne from the conversion to biodegradable materials and reusable cutleries.[1109] Additionally, as they cater to the wider community, changing the menus of children who tend to be picky eaters can lead to a lot of resistance towards healthier eating habits. Hence, there is a need to invest in incorporating health and nutrition education into mainstream curriculum.[1110]

Over the course of the 5-year programme, the Cariplo foundation had allocated €150,000 per year to cover the technical support of the Milan Food Policy.[1111] The Comune di Milano provides the personnel to manage

[1105] *Ibid.*

[1106] Milano Ristorazione (9 June 2022). *Op. cit.*

[1107] *Ibid.*

[1108] Notarangelo (9 June 2022). *Op. cit.*

[1109] *Ibid.*

[1110] *Ibid.*

[1111] C40 Cities Climate Leadership Group, Nordic Sustainability (October 2019). *Op. cit.*

this initiative and the team continues to develop new and updated food policies. By the end of 2021, the MFP team was promoted from a municipal office to an official department in charge of several key innovations and initiatives such as the following:

- **The Food Waste Hub**: In collaboration with public and private sectors, and university researchers, the hub plans on transforming the city by establishing more food hubs to provide all citizens with equal access to food, improving food waste donations and coordinating agricultural districts to secure local short supply chains for public food procurement.[1112]
- **Milano Ristorazione**: €100 million has been allocated by the state for the school canteen project (Milano Ristorazione), which offers healthy, good, educational, and fair meals to children from kindergartens to middle schools in municipality schools, private schools, and summer camps.
- **Research and Development**: MFP and MUFPP are currently working with Politecnico Milano to develop datasets and monitoring frameworks for local key performance indicators on urban food sustainability. Success would be measured by quantifiable goals such as the reduction of food waste, reduction of CO_2 emissions, and number of supporting partners.[1113]
- **Tax reduction when donating surplus food**: Milan introduced a 20% waste tax break in 2018 for food businesses donating their surplus food to charities and food banks. The municipality created a cross-departmental working group covering environmental, fiscal, and food policy staff to implement the regulation.[1114] However, it was noted that large companies would deem this amount insufficient to cover the inconvenience caused by managing the additional logistics. Recommendations to rectify this include further laws on food donations to encourage companies to respond positively, e.g., through additional incentives on top of increased tax breaks.[1115]

[1112] Porreca (9 June 2022). *Op. cit.*

[1113] *Ibid.*

[1114] C40 Cities (2019). *Op. cit.*

[1115] Cosentino, M. (6 June 2022). Banco Alimentare Food Waste Hub.

- **Shortening supply chains and stimulating local procurement**: The Distretto Agricolo Milanese (DAM) works together with the Municipality of Milan to facilitate and increase the procurement of locally sourced food. One pilot project with Milano Ristorazione for school canteens includes local procurement of rice produced by DAM in 2019. Nineteen other horticultural supply chains will be connected to locally produced food. The shortening of the production chain will be supported by the creation of new gardens for both educational purposes and local food production. This benefits the local economy and cuts down emissions by decreasing transportation.[1116]
- **Food Policy Hot Pot**: was launched in 2019 by the Municipality of Milan and Fondazione Cariplo to encourage local innovation in business models, goods, and services contributing to the Milan Food Policy. This brings together established private sector companies with 20 small- and medium-sized enterprises selected through an open call, who collaborate to co-develop projects focusing on new solutions to recover food waste, recycling methods, green procurement and supplier relationships, and technology for agriculture.[1117]

As a result of these initiatives, impacts can be seen regionally where Milan shares its food system findings and innovations. Bergamo, Brescia, and Sondrio have since launched their own food waste initiatives. Milan has also communicated its best food waste reduction practices on an international stage through the Milan Urban Food Policy Pact and the Eurocities Working Group on Food.[1118] By 2019, Milan had managed to improve resource use by significantly boosting the collection of domestic biowaste, going from 5.3% in 2011 to 20.7% in 2018. The total biowaste collection amounts to 138,000 tonnes per year, reducing annual CO_2e emissions by 41,400 tonnes. This collection has created 7.4 million m^3 of biofuel, 3.4 MWh of electrical and thermal energy, and 20,700 tonnes of high-quality organic compost.[1119] Wider benefits include the following:

[1116] *Ibid.*

[1117] *Ibid.*

[1118] Gordon & Bidault (2019). *Op. cit.*

[1119] C40 Cities. (2019). *Op. cit.*

- **Social**: With the total commercial value of surplus food donation estimated at $48.6 million, a tremendous quantity of food has been redistributed to the poorer citizens of Milan. Around 14 million meals have been thought to have been donated to Milanese food banks.
- **Health**: The redistribution of 14 million meals to food banks across this dense city has meant that many citizens may now have a nutritionally denser and perhaps healthier diet as compared to before the Milan Food Policy programme was introduced in 2016.
- **Economic**: In the Lombardy region of Italy, the standard cost for each tonne of food waste management is €90. This means that the net yearly financial savings estimate stands at around €6.3 million.
- **Environmental**: The Milan Food Policy programme has had far-reaching environmental impacts. Milan estimates that 14,500 tonnes of yearly CO_2e emissions have been reduced, while the surplus food donations have saved 5.9 billion litres of water and 4,927 hectares of land that were used in the production of that almost-wasted food.

12.4.4 *Milan Urban Food Policy Pact*

The City of Milan was the first Italian city to research food issues and consequently implement its urban Food Policy. Food is used as an entry point to sustainability, addressed through public procurement and food waste. The City of Milan also collaborates with other international organisations, including Eurocities, European Commission, and C40. The goal of the network is to create quality engagement rather than focus on the quantity of members. The EU Commission funds urban food policies through a 7-year programme of funding with several million euros committed to pilot experimental food policies as a preamble of city funding using the MUFPP framework.[1120] MUFPP also has the influence to facilitate large-scale procurement of good, healthy, and sustainable food by preventing large-scale deforestation, and procuring organic and seasonal goods in Italy. It mobilises industries with the purchasing power to shift the local market.

[1120] *Ibid.*

The secretariat facilitates knowledge exchanges, hosts international delegations, and catalyses urban food policy exchange through different cities, especially in South America and Africa.[1121] MUFPP is currently organising regional forums in the Asia Pacific and North America. They are also developing a fellowship programme in 2023 to accelerate food policy through webinars and field visits to maintain its vision and primary role through continuous re-innovation. This is also facilitated through knowledge exchange from signature cities with events, workshops, and forums which MUFPP organises at least once every 6 months. By keeping Milan's local policy at the forefront MUFPP aims to retain the credibility of Milan's food policy and feed the international knowledge exchange.[1122] Within this network of cities, the urban food sustainability agenda has been advanced through urban food policies and innovation on the local scale.

Some key challenges identified include language barriers as the MUFPP team is limited to French, English, Spanish and Italian. However, this limits outreach to APAC countries that do not primarily communicate in these languages. Key entry points to successful corporation were due to identifying food policies which would be relevant to local contexts, political priorities, and the urban environment of specific cities. Once collaborations have been initiated, things move more efficiently due to international cooperation and shared interests. Due to consistent efforts between 2020 and 2021, this has led to the development of an organisational/divisional forum in APAC, which Indonesia hosted in Bandung, August 2022. APAC has also organised a food policy training programme for city officers, with a successful sign-up rate of over 100 participants due to mutual interest in the potential impact of this initiative.[1123]

COVID-19 deeply affected operations in Milan as Italy was unprepared for the crisis, leading to an emergency halt on all innovations and initiatives at that time (public orchards, financial tools, events, forums, awards, etc.). The MFP and MUFPP team pivoted their efforts towards food procurement and accessibility to meet citizen needs during the crisis.

[1121] *Ibid.*

[1122] *Ibid.*

[1123] *Ibid.*

The team developed the Food Aid System during the pandemic in the day, while at night, they continued to coordinate regional cities to manage the lockdown and ensure continued local food supply. Information was shared between cities to establish the mechanism of food aid.

Resultingly, the Milan Pact Talks was created which consisted of 100+ videos, comprising a collection of first-response videos from mayors and officers. This helped Italy to collectively manage the COVID-19 aftershock on the food system, reorganise markets, and establish a food aid system for citizens to sustain the food economy (predominantly farmers and restaurants). Another key role was to provide healthy food for people at home and give hope and support to public officers.[1124]

This successfully mitigated the impact on food wholesale markets which initially faced some logistics issues in the first month of the pandemic. Fruits and vegetable sales were unaffected, but seafood sales faced issues as people did not cook seafood at home as much. The food economy was rediverted as retailers and supermarkets would buy from the wholesale market and sell it to customers back home, replacing the original demand from restaurants.[1125] However, one prevailing problem is that delivery services had grown a lot during COVID which generated more waste due to the need for single-use plastics.[1126]

MUFPP has positively impacted regional and international knowledge on food waste and urban sustainability by developing a set of indicators alongside Milan Polytechnic to kickstart the evaluation of impacts on the Milan food system.[1127] This set of indicators consists of 37 recommended approaches which analyse food sustainability with a large influence on citizen life.[1128] The resulting vision is to sensitise governments on the role of cities to drive food system change and advocate the role of cities to practice sustainability, as discussed in the UN City and Local Food System Pre Summit 2021. This provides stimulus to local authorities of the metropolitan

---

[1124] Gavazzeni (7 June 2022). *Op. cit.*

[1125] Zaffra, N. (7 June 2022). Foody Wholesale Market, Operated by SOGEMI.

[1126] Calori, A. (8 June 2022). ESTA Economia e Sostenibilita.

[1127] Urban Sustainability Exchange. (n.d.). *The Milan Food Policy.* https://use.metropolis.org/case-studies/the-milan-food-policy.

[1128] Gavazzeni (7 June 2022). *Op. cit.*

area and within their international network to follow the example of Milan in working toward a more sustainable approach to their food systems.

The Milan Pact Awards is another initiative established as a tool to transfer innovative solutions between cities. Good practices formally submitted by Mayors every 2 years build a library of knowledge, which inform the level of impact and give the international jury of experts panel a glimpse of the joint impact of their respective food policies. Impact is assessed through the 37 measures of the SDG framework (self-assessment) or the Food and Agriculture Organisation's (FAO) 44 indicators to monitor cities. The Milan Pact Awards and the APAC Food Policy training platform allows for the cross pollination of international knowledge and can guide other cities on how to achieve their goals.[1129]

To ensure the longevity and transferability of these programmes, the MUFPP now goes by regional elections to select the steering committee from two cities per continent where they would have to serve a 2-year mandate.[1130] These cities oversee contribution to shaping new global role of cities that stand out as front runners for sustainable development through the localisation of SDGs.[1131] Challenges were also overcome due to the close working partnership with the municipal governments and external organisations as political commitments and politician visions were important to steer, staff, and fund the programme.[1132] MUFPP, through its library of knowledge, sheds light on challenges faced by cities around the world. These challenges encompass overly ambitious goals, challenging political situations, a lack of capacity, insufficient participation from critical actors, conflicting national policies, and inadequate engagement mechanisms which are impeding long-term sustainable development.

However, the experiences shared through MUFPP's international outreach also reveal a roadmap for cities to overcome these obstacles and forge a path towards more sustainable and effective urban food policies.

[1129] *Ibid.*

[1130] *Ibid.*

[1131] Milan Urban Food Policy Pact. (15 April 2021). *Results of the Election of the 3rd MUFPP Steering Committee.* https://www.milanurbanfoodpolicypact.org/results-of-the-election-of-the-3rd-mufpp-steering-committee/.

[1132] Porreca (9 June 2022). *Op. cit.*

Cities can adopt a combination of strategies to navigate these challenges successfully:

- **Strategic entry points**: Cities should carefully identify and select entry points that are not only likely to yield success but also demonstrate tangible impacts. By doing so, they can build a coalition of support across government bodies and other stakeholders, ensuring the longevity of their initiatives.
- **Long-term planning**: To circumvent the constraints imposed by political cycles and elections, cities should adopt longer planning cycles. This approach enables continuity and stability in food policy implementation, reducing vulnerability to changes in leadership.
- **Multi-sector engagement**: Establishing mechanisms for cross-cutting agency collaboration and multi-sector engagement, such as food councils and inter-agency working groups, is crucial. These platforms should include the active and meaningful participation of civil society, enhancing the diversity of perspectives and expertise.
- **Securing governmental support**: Cities should seek support not only at the local level but also from subnational regional or national governments. Collaborative efforts that span multiple levels of government can help address jurisdictional conflicts and create a more enabling environment for urban food policies.
- **Learning networks**: Building demand-driven and mutually beneficial learning and exchange networks with counterpart local authorities and practitioner communities can facilitate the sharing of best practices and innovative solutions. These networks can accelerate progress and foster a culture of continuous improvement.

12.5 Conclusion

Milan has assiduously promoted resource conservation and the development of a circular economy through food waste management, scaling the effort to other cities in Italy and other countries. The positioning of the Milan Food Policy initiative within the Mayor's Office and directly under the mandate of the first Vice Mayor ensures its visibility and acknowledgement. To overcome the challenges of a siloed government system and promote international corporation towards environmental sustainability, the MFP and MUFPP use food as an entry point for joint sustainable transformation of growing cities. Within the MFP and MUFPP, initiatives have had lasting impacts on cities around the world through knowledge sharing and facilitation of sustainable innovations.

MFP has made significant progress in monitoring, evaluating, and tackling food waste mainly through promoting local procurement, with the support of a strong local food infrastructure and good organisation. Through initiatives like the Food Waste Hub and Milano Ristorazione, the city aims to ensure equitable access to food while reducing CO_2 emissions and supporting local economies. Despite challenges such as insufficient tax incentives for large corporations, Milan's efforts have garnered regional and international recognition, leading by example. By 2019, Milan had achieved quantifiable results, including a substantial increase in biowaste collection and significant reductions in CO_2 emissions. These achievements not only address environmental concerns but also deliver social, health, and economic benefits to the community.

MUFPP is an international leader in sustainable urban development and waste mitigation. Through initiatives like the Milan Pact Awards and strategic partnerships with institutions like Milan Polytechnic, the city has pioneered a comprehensive set of indicators to assess food sustainability and foster international knowledge exchange. Overcoming challenges through close collaboration with municipal governments and external organisations, Milan's approach emphasises long-term planning and multi-sector engagement, backed by strong governmental support. By sharing experiences and best practices globally, Milan is paving the way for cities worldwide to navigate similar challenges and advance toward more sustainable and effective urban food policies, ultimately contributing to the realisation of the Sustainable Development Goals on a local level.

In summary, while there are challenges in developing and implementing urban food policies, Milan has strategically approached their goals, adopting long-term planning, promoting collaboration, securing support at various government levels, and fostering learning networks; cities can forge a path toward healthier, more sustainable, and equitable food systems that benefit their residents and contribute to global food security.

EPILOGUE

Building a smart city is a journey, not a one-off activity. As the 11 case cities demonstrate, it often starts with a desire and vision to improve the city's economy, society, and/or environment. Smart city initiatives can take many forms and scales, from citywide digital infrastructure like free public Wi-Fi to innovation hubs, sensors in parks, and digital signages on public transit to individual-level devices like robots, telemedicine, e-wallets, and mobile apps. Even though smart city implementation is often motivated by technology advancements (to harness the latest capability), its success necessitates the consideration and partnership with people and organisations to develop an enabling ecosystem that supports daily life and routines. Challenges are inevitable (e.g., in terms of resources, awareness, and utilisation) but 10 Cs have emerged as critical to people-centric smart city development:

- **Community engagement and evaluation**: Being able to understand and address the diverse and changing needs and demands of communities allows cities to propose innovations with positive and relevant community impact. Many cities in this book carry out regular data collection (from surveys to sensors and open data platforms) and community engagement activities through various traditional (e.g., email updates and feedback surveys), real-time (e.g., civic engagement apps), and participatory strategies (e.g., co-design workshops, public hearings, and town halls) to learn and respond to citizens' needs. Boston's MONUM digital placemaking innovations highlight the importance of public engagement and evaluation in understanding on-the-ground issues in the neighbourhoods to advance social inclusion. Seoul's I Will Centres demonstrate community engagement by integrating the concerns, expertise, and insights of a wide range of stakeholders in addressing internet addiction among youth. Besides working with schools to break the addiction, Seoul also follows up with recovery environment creation at home in a multi-pronged approach of creating long-term assessment frameworks and leveraging community support in combating internet addiction.

- **Continuous adaptation and purpose-driven innovation**: Tech and innovation industries have progressed through research and knowledge development and exchange to evolve and remain relevant. Many cities and countries use their most pressing urban issues like climate resilience, over-tourism, and ageing population, to catalyse and establish purposeful innovations and smart cities. Japan's fast-ageing population has fuelled age-tech innovations to help people live better as they age and to relieve labour shortages and care stress burden. Copenhagen's Climate Resilient Neighbourhood initiative is the city's response to global climate action, combining nature-based solutions with urban planning to enhance liveability and resilience in flood management. Melbourne's urban greening efforts likewise prioritise the development of green infrastructure for climate resilience, continuously adapting for environmental sustainability. Milan's focus on research and development, along with monitoring frameworks for urban food sustainability, entrenches data-driven decision-making and continuous improvement. Learning from setbacks is integral to the innovation process. Auckland, Gothenburg, and Vancouver advocate for experimentation, strategic planning, and iterative learning as key strategies in navigating uncertainties to meet evolving needs and expectations. Pilot projects are implemented to test new technologies and solutions before scaling up. Melbourne's Argyle Square smart park pilot project adopts a problem-driven approach, focusing on human-centric design and big data analytics to address climate challenges and make public open spaces comfortable for communities to use.

- **Commitment and cross-sector collaboration**: A functional policy framework that overcomes silos and fosters greater cross-functional collaboration across the entire public sector is fundamental to the efficacy of smart city development. Several dimensions of policy integration are suggested to bring together all the factors, fundamentals, and layers that support a smart city, e.g., aligning and linking up national, regional, and local development, strengthening horizontal integration across sectors and disciplines, promoting vertical integration across different government levels, and encouraging a whole-of-society approach to forge new ways to smart city goals with the engagement of people, private, and public sectors.[1133] These integration and linkages are easier said than done, requiring commitment and perseverance. Several approaches and strategies prevail. Boston's MBTA convenes an in-house, cross-functional team to ensure capabilities for technology innovations towards a more inclusive, accessible transit system. Vancouver's approach to appointing a Chief Technology Officer further highlights the importance of having an autonomous department empowered to lead the implementation of the digital strategy, including to enact the city's vision and manage cross-sector communication, coordination, and collaboration. Milan's partnership-building efforts involve a central Food Policy and an administrative setup, the Urban Food Policy Pact secretariat, to drive cross-sector collaboration with local non-profit, agricultural industries, as well as cross-city collaboration — other cities in Italy and elsewhere with similar visions. Collaboration network enables peer-to-peer learning, offering a platform for communities of practice to form, for members to share knowledge and best practices.
- **Collaborative regulatory frameworks**: Innovation often requires not only supportive policy but also people-centred regulatory frameworks to flourish. This could range from introducing new regulations to modifying existing, outdated regulations, to support digital transformation and smart city development. As Auckland demonstrated, the success of innovation hubs hinges on effective operational

---

[1133] Yuen *et al.* (2024). *Op. cit.*

leadership, commitment, and policy framework. Tokyo's policy-driven governance framework showcases strong leadership and strategic vision essential for promoting digital literacy and responsible technology use. Melbourne's planning scheme amendment on Open Space Contribution Framework (a mandatory requirement for private sector developments) and the recently proposed amendment on Sustainable Building Design (a minimum Green Factor score of 0.55) exemplify the importance of actively pursuing changes to planning regulations to enable the implementation of urban green initiatives.

- **Capacity building and talent acquisition**: Realistic implementation plans coupled with a long-term plan for technological and workforce capacity building are essential to smart city building. A clear understanding of technology capacity and having the talent to implement it is key to minimising the expectation-deliverable gap (avoid overselling what the technology can do). Auckland's Tech Tāmaki Makaurau strategy has developed a series of actions to address the need for creating a sustainable and equitable tech ecosystem with home-grown and global talent. Hangzhou builds on its home industry, Alibaba's expertise, further underscores the importance of building local capacity to support new digital initiatives. Both Hangzhou and Vancouver have established long-term capacity building by engaging local universities and design institutions to contribute to their workforce development. Seoul's investment in specialist training and professional counselling builds local capacity in targeted fields to address societal challenges.
- **Communication and buy-in**: Effective communications are critical for building trust and gaining support for smart city development. This applies to both internal communications within government sectors and with external collaborations and the community. Copenhagen champions partnerships among government departments maintaining good communication with city leaders for citywide implementation. A critical factor to Auckland's digital transformation journey is obtaining buy-in from top levels of the government, other agencies, business partners, and the public, upon verifying the value and feasibility of digital innovations. Guangzhou's efforts to bridge the digital divide and Boston's community-centric approach further

showcase how to identify stakeholders, interact with them, and maintain communication throughout the innovation and implementation process. These cities highlight the importance of securing buy-in from various stakeholders, including government agencies, businesses, and communities.

- **Capital investment and co-funding**: Accessing and securing adequate funding are crucial for sustaining innovation efforts. Funding arrangements range from government funding schemes (subsidies, incentives, and tax benefits), private investment, and co-funding mechanisms (cash and land) from key stakeholders, including the public sector, developers, businesses, and the wider community. Hangzhou's government investment schemes are one example of a top-down approach, whereas Vancouver's partner ecosystem approach enables the scaling of smart governance initiatives while managing procurement and development challenges from private partnerships. Tokyo's interdisciplinary collaboration and subsidies demonstrate how governments can overcome financial barriers to adopting next-generation robots. Milan's Food Policy and Melbourne's green infrastructure projects both adopt co-funding mechanisms by securing funding from public and private sectors to support their initiatives.
- **Citizen awareness and ownership**: Strengthening citizen knowledge and participation is key to providing equitable access to sustainability transformations. A variety of channels can be used to cultivate community interest and ownership in smart city projects, such as community events, forums, workshops, how-to guides, citizen science, and volunteers. Gothenburg's strategies, such as urban experimentation and citizen engagement platforms, demonstrate efforts to bridge the knowledge-awareness gap. Melbourne's Canopy Green Roof Forum facilitates a knowledge-sharing platform among various stakeholders, such as building owners, built environment professionals, researchers, and developers, to share industry best practices in green infrastructure and to exchange ideas to overcome challenges in implementing urban greening initiatives. Milan's approach reaches out to citizens through education programmes and public engagement activities to ensure community ownership of sustainable food policies. Seoul goes even further by addressing the social reluctance of vulnerable youths and

their families to participate in digital youth empowerment. The city pinpointed pain points, such as social stigma related to internet addiction and, in response, marketed their programmes as fun outdoor activities and skills training to cast a positive light in empowering their youths.

- **Clarity in objectives and target beneficiary**: Smart city development is a complex endeavour, involving many actors and factors. Having a well-defined and clear vision, objectives, and target beneficiaries is an important first step of a smart city. As illustrated by the 11 case cities, there is a multiplicity of smart city objectives, from economic diversification to improving quality of life, social inclusion, participatory governance, responsible use of resources, and building climate resilience. A common thread is a desire to improve the city and its communities. Identification of target beneficiaries (both direct and indirect) and stakeholders, and a sound understanding of their problems and experiences are vital for the development and impact of a people-centric smart city. An increasing number of cities, e.g., Boston and Melbourne, have included the beneficiaries (user needs and expectations) as part of their social innovation process, enhancing user experience and optimising digital platforms while fostering engagement. Hangzhou's policy guidance and clear definition of what hybrid retail entails guide both public and private sectors' technological and policy-based development to have cohesive progression. Gothenburg's initiatives to further improve Min Stad through visual prioritisation and digital twins highlight the response to user feedback and the importance of user-centric, easy-to-use digital design and interfaces in smart tools.
- **Consistency in implementation and maintenance**: Ensuring consistency in implementation and maintenance is essential for long-term success. This can be achieved through long-term planning and adaptive measures. Melbourne's smart environment actions align with the city's long-term climate change mitigation goals for 2050, ensuring that measures continuously and effectively address climate risks. Guangzhou's focus on platform optimisation and Boston's emphasis on cross-department collaboration highlight strategies for sustaining the implementation and maintenance of digital initiatives.

Vancouver's people-first strategy keeps the transformation process accountable by focusing on tangible outcomes, ensuring that smart governance initiatives resonate with citizens' needs and expectations. As Auckland demonstrated, the success of innovation hubs hinges on effective operational leadership, commitment, and policy framework. Unforeseen events are an inevitable part of long-term development, and it may not always be possible to prevent them totally. It is important for cities to develop the adaptive capacity to navigate and deal with unexpected challenges and risks when they occur. Milan's response to the COVID-19 pandemic demonstrated the importance of adaptive measures and crisis management in ensuring food security and resilience. The city's swift action and existing collaboration with regional cities mitigated the impact on the food system and supported local economies during the pandemic and unexpected circumstances.

www.ingramcontent.com/pod-product-compliance
Lightning Source LLC
Chambersburg PA
CBHW070757100725
29074CB00020B/56